Adaptation and Impacts of Climate Change

Adaptation and Impacts of Climate Change

Editor: Mary D'souza

MURPHY & MOORE
www.murphy-moorepublishing.com

www.murphy-moorepublishing.com

Ⓜ MURPHY & MOORE

Cataloging-in-publication Data

Adaptation and impacts of climate change / edited by Mary D'souza.
   p. cm.
Includes bibliographical references and index.
ISBN 978-1-63987-692-1
1. Climatic changes. 2. Adaptation (Biology). 3. Human beings--Effect of climate on.
4. Climatology. I. D'souza, Mary.
QC903 .A33 2023
363.738 74--dc23

© Murphy & Moore Publishing, 2023

Murphy & Moore Publishing
1 Rockefeller Plaza,
New York City,
NY 10020, USA

ISBN 978-1-63987-692-1

able of Contents

Preface VII

Chapter 1   **Role of Small Grains in Adapting to Climate Change: Zvishavane District**..1
Tendai Nciizah, Elinah Nciizah, Caroline Mubekaphi and
Adornis D. Nciizah

Chapter 2   **Use and Impact of Artificial Intelligence on Climate Change Adaptation**.. 20
Isaac Rutenberg, Arthur Gwagwa and Melissa Omino

Chapter 3   **Differential Impact of Land Use Types on Soil Productivity Components in Two Agro-Ecological Zones**.................................... 40
Folasade Mary Owoade, Samuel Godfried Kwasi Adiku,
Christopher John Atkinson and Dilys Sefakor MacCarthy

Chapter 4   **Women Participation in Farmer Managed Natural Regeneration for Climate Resilience: Laisamis, Marsabit County**.........................53
Irene Ojuok and Tharcisse Ndayizigiye

Chapter 5   **Adaptive Capacity to Mitigate Climate Variability and Food Insecurity of Rural Communities Along River Tana Basin**...........................72
David Karienye and Joseph Macharia

Chapter 6   **Biomass Burning Effects on the Climate over Southern West Africa During the Summer Monsoon**...84
Alima Dajuma, Siélé Silué, Kehinde O. Ogunjobi, Heike Vogel,
Evelyne Touré N'Datchoh, Véronique Yoboué, Arona Diedhiou
and Bernhard Vogel

Chapter 7   **Multifunctional Landscape Transformation of Urban Idle Spaces for Climate Resilience**..102
David O. Yawson, Michael O. Adu, Paul A. Asare and
Frederick A. Armah

Chapter 8   **Economic Analysis of Climate-Smart Agriculture Technologies in Maize Production in Smallholder Farming Systems**.................................128
Angeline Mujeyi and Maxwell Mudhara

Chapter 9   **Pyrolysis Bio-oil and Bio-char Production from Firewood Tree Species for Energy and Carbon Storage in Rural Wooden Houses**...143
Miftah F. Kedir

Chapter 10   **Constraints to Farmers' Choice of Climate Change Adaptation Strategies in Ondo State**..16
George Olanrewaju Ige, Oluwole Matthew Akinnagbe,
Olalekan Olamigoke Odefadehan and Opeyemi Peter Ogunbusuyi

Chapter 11   **Climate Change Resistant Energy Sources for Global Adaptation**..........................17
Oluwatobi Ololade Ife-Adediran and Oluyemi Bright Aboyewa

Chapter 12   **Building Livelihoods Resilience in the Face of Climate Change: Case Study of Small-Holder Farmers**..18
Saumu Ibrahim Mwasha and Zoe Robinson

Chapter 13   **Intangible and Indirect Costs of Adaptation to Climate Variability Among Maize Farmers: Chirumanzu District**..20
Dumisani Shoko Kori, Joseph Francis and Jethro Zuwarimwe

                 Permissions

                 List of Contributors

                 Index

Preface

ng-term changes in temperature and weather patterns are referred to as climate change.
ese changes could be due to natural causes such as oscillations in the solar cycle. However,
man activities are the primary cause of climate change owing to the burning of fossil fuels
ch as gas, coal and oil. The combustion of fossil fuels produces greenhouse gas emissions,
hich act like a blanket wrapped around the Earth, trapping the sun's heat and raising the
mperature of Earth. Methane and carbon dioxide are the two examples of greenhouse gas
nissions that contribute to climate change. Adaptation allows humans to prepare for some
the most expected repercussions of climate change by limiting their negative effects on
osystems and human well-being. It involves better disaster planning, strengthening water
nservation programs, and building early warning systems for extreme heat events. This
ok provides comprehensive insights on the adaptation and impacts of climate change. It
ill provide comprehensive knowledge to the readers.

is book is the end result of constructive efforts and intensive research done by experts in
is field. The aim of this book is to enlighten the readers with recent information in this
ea of research. The information provided in this profound book would serve as a valuable
ference to students and researchers in this field.

the end, I would like to thank all the authors for devoting their precious time and
oviding their valuable contribution to this book. I would also like to express my gratitude
my fellow colleagues who encouraged me throughout the process.

Editor

Role of Small Grains in Adapting to Climate Change: Zvishavane District, Zimbabwe

Tendai Nciizah, Elinah Nciizah, Caroline Mubekaphi and Adornis D. Nciizah

Contents

Introduction .
Smallholder Farming in Zimbabwe
Farmers' Perceptions on Climate Change in Zvishavane
Smallholder Solutions for Climate Change
Characteristics and Impact of Small Grain Production on Climate Change Adaptation in
Zvishavane District ....
Perceived Barriers to Small Grain Production in Zvishavane
Lack of Inputs
Labor Intensiveness
The Challenge of Quela Birds
Low Yields
Lack of Knowledge
Lack of Markets

T. Nciizah
Department of Sociology, Rhodes University, Makhanda, South Africa

E. Nciizah
Department of Development Studies, Zvishavane Campus, Midlands State University,
Zvishavane, Zimbabwe
e-mail: nciizahe@staff.msu.ac.zw

C. Mubekaphi
School of Agricultural, Earth and Environmental Sciences, University of KwaZulu-Natal,
Scottsville, South Africa

A. D. Nciizah (✉)
Soil Science, Agricultural Research Council – Institute for Soil, Climate and Water,
Pretoria, South Africa
e-mail: nciizaha@arc.agric.za

Low Extension Services and Government Support .
Food Preferences
Interventions to Overcome the Barriers to Small Grains Adoption
Conclusions ....
References

Abstract

Climate change has become one of the most profound threats to smallholder agriculture in semi-arid and arid areas. Farmers in this sector are especially vulnerable to climate change due to reliance on rain-fed agriculture, limited access to capital and technology among other challenges. While several potential adaptation options exist, many barriers hinder effective adoption of these practices, hence production in marginal areas remains very low. This chapter discusses crop adaptation through the adoption of small grains in Zvishavane rural, a semi-arid area in Zimbabwe. Small grains are conducive in hot areas; their drought-tolerant nature enables them to thrive in marginal areas making them an appropriate strategy in responding to climate change. However, several production and policy challenges associated with small grain production hinder their adoption by farmers. In view of this, this chapter discusses the potential of small grains as an adaptation strategy to climate change in Zvishavane District, Zimbabwe, and addresses potential challenges and opportunities for increased adoption and future research. The review showed that farmers in Zvishavane have perceived climate change due to noticeable changes in rainfall and temperature patterns in the past years. Despite small grain production being the best strategy due to drought and high temperature tolerance, an insignificant number of Zvishavane farmers is involved in small grain production. This is due to numerous barriers such as high labor demand associated with small grain production, the challenge posed by the quelea birds, food preferences, low markets, and low extension services and government support. It is therefore necessary to encourage adoption of small grains by developing improved varieties, adoption of climate smart agricultural practices, improved technical support, and access to markets among other interventions.

Keywords

Adaptation · Climate change · Mitigation semi-arid areas · Small grain production · Smallholder farmers · Vulnerability

Introduction

Climate change has become one of the major threats to rain-fed agriculture resulting in major drawbacks in agricultural production and food security. IPCC (2007) notes that climate change has significantly modified rainfall and temperature patterns in many parts of the world. There is strong evidence of erratic rainfall and a noticeable shortening of the agricultural seasons worsened by higher ambient temperatures resulting in perennial crop failure and frequent droughts across the world (Jerie and

Ndabaningi 2011). The adverse effects of climate change are more pronounced within the smallholder agricultural sector in Africa, including Zimbabwe, compared to the more technologically advanced and high resource input commercial sector. Generally, smallholder agriculture is practiced by families or households using only, or mostly, family labor and deriving from that work a large but variable share of their income, in kind or in cash (HLPE 2013). It therefore follows that the livelihoods of smallholder farmers largely depend on agriculture and are particularly vulnerable to several hazards. Consequently, if changing climatic conditions continue unabated, traditional rain-fed agricultural systems will become increasingly unsustainable with severe food security implications.

Temperature in the sub-Saharan Africa region is predicted to increase by between 1 °C and 2.5 °C by 2030, which will ultimately result in yield losses of up to 20% by the 2050s depending on region (IPCC 2007). Consistent with regional data, Zimbabwe's annual mean surface temperature has warmed by about 0.4 °C from 1900 to 2000 with the period from 1980 to date being the warmest on record. Moreover, the timing and amount of rainfall received are becoming increasingly uncertain, with current data showing that during the last 30 years there has been frequent droughts and rainfall has significantly decreased with shorter seasons (GoZ 2014). Consequently, the Food and Agriculture Organization (FAO 2007) observed that the crop failures have been a result of early termination of the rains in most seasons or low rainfalls in Zimbabwe, especially in semi-arid areas. Gukurume (2013) perceived that climate change in countries like Zimbabwe has presented insurmountable challenges to the agricultural sector as well as agricultural sustain-ability. Droughts, floods, increased temperature, increased rainfall variability, and declining precipitation negatively affected agriculture in Zimbabwe, with other districts recording almost nothing in terms of output (Zimbabwe Agricultural Sector survey 2019). The 2018/19 cropping season was marked by a late onset of rains, which resulted in abnormally dry conditions, affecting agricultural activities such as land preparation and planting (IPC Zimbabwe 2019).

The effects of climate change in Zimbabwe are exacerbated by the concentration of most smallholder farmers in marginal soils in semi-arid and arid areas located in the natural region IV and V, which makes them more vulnerable to climate change due to their low adaptive capacity. High temperature and erratic rainfall have resulted in constant low yields of the preferred maize staple in the country. Maize is not a drought-tolerant crop; hence, it is highly sensitive to harsh environmental condi-tions, which negatively affect the yield (Muzerengi and Tirivangasi 2019). There is therefore a need to implement adaptation and mitigation strategies that will ensure resilience to climate change. Adaptation has the potential to reduce adverse impacts of climate change and to enhance beneficial impacts and hence reduce vulnerability both in the short and long term (IPCC 2001). There are several adaptation interven-tions that have been suggested for smallholder farmers including but not limited to crop relocation, adjustment of planting dates, and crop variety; improved land management, e.g., erosion control and soil protection incorporating agroforestry (IPCC 2007). However, due to the vulnerability of maize to drought, there is the need to move away from the production of maize and look for alternative crops that thrive under changing climatic conditions. Sharma et al. (2002) highlighted that there is an urgent need to focus on improving crops relevant to the smallholder farmers and poor consumers in the semi-arid areas. This chapter focuses on the adoption of small grain crops, which have been shown to thrive under harsh conditions, as one of the most viable strategies that must be consistently implemented in such semi-arid and arid areas as Zvishavane District in Zimbabwe, which are facing significant production challenges due to climate change. Zvishavane District has experienced recurrent droughts, inconsistence rainfall, and is consequently one of the districts that constantly faces food insecurity in the country. According to Mawere et al. (2013), for the past two decades, Zvishavane, among others in Zimbabwe and beyond, has witnessed pronounced increases in temperature, recurrent droughts, and unpredictable rainfall patterns, yet people mainly depend on rain-fed agriculture and natural resources for their livelihoods. This chapter discusses the potential of small grains as an adaptation strategy to climate change in Zvishavane District, Zimbabwe, and addresses potential challenges and opportunities for increased adoption and future research. The chapter will answer the following questions:

(i)   What are the Zvishavane farmers' perceptions on climate change?
(ii) To what extent are farmers in Zvishavane adopting small grains?
(iii) What are potential barriers/challenges and opportunities for increased adoption of small grain crops in Zvishavane District?

Smallholder Farming in Zimbabwe

The smallholder sector is known for its small farms, mostly less than 2 hectares that are labor-intensive, uses traditional production techniques, and often lack institutional capacity and support (Louw 2013). The definition of smallholder farmer depends on the country; for instance, in Zimbabwe, smallholder farmers are typically farmers who are food insecure, predominantly in low rainfall areas who depend on rain-fed agriculture. Their reliance on rainfall for agricultural production exposes them to the risks posed by climate change. Smallholder farmers face challenges such as lack of infrastructure, reliance on rain-fed systems, highly degraded soils, a variety of chronic crop-affecting diseases, tough economic conditions due to poorly functioning markets, and limited access to credit, knowledge, and lack of farming skills (Mutekwa 2009). Lack of assets, information, and access to services deters smallholder farmers' involvement in possibly profitable markets. Moreover, most smallholder farmers are confronted with labor limitations as a result of HIV/AIDS, chronic illness, poverty, and illiteracy. Furthermore, smallholder farmers lack adequate knowledge, which leads to poor selection of cultivars, fertilizer application, and delayed planting of crops. As a result, the crop yields are negatively impacted. Worse still they are far from road networks to transport inputs, produce, and also access information and have a tendency to use inefficient modes of transport such as animals. In Zimbabwe there is nonexistence of investment in infrastructure, for instance, roads, storage, and market facilities and this handicap the potential role of smallholder farmers. This problem is not only common in Zimbabwe but the rest of Africa as well as most African government policies are unsuitable and inconsistent, and do not provide an enabling environment for the development of the small grains sector.

Mutasa (2011) highlighted the reasons below as challenges or factors that farmers are faced with that also increase farmers' vulnerability to climate change.

(a) Poor soil moisture holding capacity
(b) Lack of necessary farming knowledge
(c) Lack of draught power
(d) Difficulties accessing the appropriate inputs on time
(e) Inadequate farming space
(f) Poor rains and/poor rainfall distribution
(g) Communities' failure to fully recover from previous drought disasters
(h) Poor fiscal policies resulting in hyperinflation and inaccessibility to cash especially after the dollarization of the economy
(i) Poor governance evidenced by the suspension of humanitarian organizations' operations when they were greatly needed
(j) Politicization of food assistance and input schemes (benefitting only supporters of a particular party)
(k) Corruption associated with input facilities
(l) Communities' focus on maize production at the expense of other traditional crops
(m) Depletion of human resources and sources of income due to brain drain, HIV, and AIDS and other epidemics

All the above mentioned challenges affect their competitiveness and it leads to them having high transport costs to sell their products in the market and buy inputs. In addition, smallholder farmers face challenges in accessing markets due to inadequate infrastructure, low productivity levels, inconsistencies in supply, and low quality owing to poor post-harvest practices (FAO). Smallholder farmers do not make much income as they lack a reliable market and as a result sell their produce at their farm gates or local market where they sell below market value (Aaron 2012). They tend to be inconsistent in production due to lack of bargaining power and inability to reliably supply their products fresh to markets (Aaron 2012). Harvested crops are threatened by insects such as weevils, larvae, termites, and rodents such as rats; these insects may be present in the grains making it less marketable. Some of the pests are attributed to lack of storage facilities as some smallholder farmers store grain in their sleeping quarters whereas those who have structures tend to be in very poor conditions and hence cannot hinder pest attack.

Changing climatic conditions have become one of the major challenges affecting smallholder farmers in Zimbabwe and the rest of the world. In 2016 and 2017, Zimbabwe had a drought, which was aggravated by the impact of El Niño. Smallholder farmers tend to suffer from these disasters due to their low adaptive capacity, which increase their vulnerability, hence they have no means of recovering from the effect this has on their livelihoods. Most smallholder farmers have a tendency to recover by selling their productive assets, such as their livestock or land. Moreover, these persistent droughts, which have taken place over the past few years due to changes in climate, continue to erode maize yields. For instance, maize yields have been reduced to less than 1.5 ton ha^{-1} against a potential yield of more than 5 tons ha^{-1} (de Jager et al. 2001).

Semi-arid and arid regions like Zvishavane district have some of the worst affected farmers in Zimbabwe. Zvishavane is a district in Zimbabwe bordered by Mberengwa, Chivi, and Shurugwi districts. The mean annual temperature is 20 °C although high temperatures of up to 30 °C have been recorded during the hot months from October to December (Nciizah 2014). The area receives an annual rainfall of about 450–600 mm placing it in region IV but during droughts there can be just 250 mm of rain (Oakland Institute Undated). According to the OCHA report (2012), region IV is subject to periodic seasonal droughts and severe dry spells during the rainy seasons, and crops can only be intensified by growing of drought-tolerant crops. Natural farming regions are a classification of the agricultural potential of the country, from natural region 1(>1000 mm per annum), which represents high altitude wet areas to natural farming region V, which receives low and erratic rainfall averaging 550 mm per annum (Mugabe et al. 2007). The table below shows the agroecological regions in Zimbabwe and their characteristics.

Rainfall in Zvishavane has been erratic such that in some areas like Mazvihwa, rain-fed agriculture has become unreliable, worsened by droughts that have gripped the country (Nciizah 2014). Despite the hot climatic condition of the district, agriculture remains the main source of livelihoods (Mugiya and Hofisi 2017). Just like all areas in Zimbabwe, the main crop grown by smallholder farmers in Zvishavane is maize. Droughts and little rainfall in the district have led to low maize crop yields leading to food insecurity to be high in the area. According to ZimVac (2015), Zvishavane was one of the districts with the highest food insecurity levels with 42.2%. In 2016, the level went up with 50% households being food insecure, hence the need for adaptation and mitigation measures in smallholder farming areas (ZimVac 2016). Adaptation measures include adjustment of planting dates and crop variety, crop relocation among others. However, one of the most recommended approach is the growing of drought-tolerant crops such as small grains (Gukurume 2013; Muzerengi and Tirivangasi 2019; Musara et al. 2019). Adoption of small grains becomes a crucial necessity that must be embraced by households in the area. Small grains grown in the area are sorghum and millet, with finger millet being more common than pearl millet (Nciizah 2014).

Farmers' Perceptions on Climate Change in Zvishavane

Smallholder farmers in Zvishavane perceive climate change using two variables, which are temperature and precipitation paying attention to events that have been occurring in the area for the past 10 to 20 years. Perceiving that climate is changing is crucial as it leads to adaptation. According to Jiri et al. (2015), perceptions help to shape smallholder farmers' coping and adaptation strategies. Adaptation is a two-step process, which requires that farmers first notice that the climate has changed and then secondly, implement adaptation strategies. Farmers' ability to acknowledge the importance of adapting therefore largely depends on whether they have observed that there is climate change in the first place (Nciizah 2019). Most studies have shown that farmers who perceive that the climate is changing in line with the actual climate change records are most likely to adapt to climate change (Jiri et al. 2015). Simba et al. (2012) concurs that the most crucial element in spearheading adaptation options is for farmers to perceive climate change. Such farmers who perceive that the climate has changed, for instance, will be able to realize a sharp decrease in maize yields. Numerous studies have been done on farmers' awareness on climate change (Gbetibouo 2009). A study by Okonya et al. (2013) in Uganda showed that 99% of all households interviewed had observed a change in the climate in the last 10 years.

The greatest concern for farmers in Zvishavane is the drastic changes in rainfall patterns and temperature. This is confirmed by the records from meteorological centers. A study by Jiri et al. (2015) also confirmed the decline in rainfall and an increase in temperatures over the years. Similarly, Mutekwa (2009) showed that farmers in Zvishavane were mainly concerned with precipitation and temperature-related weather events, as the ones that are a real concern in relation to their agricultural activities. There are now more experiences in dry spells and cold spells. Murowa ward in Zvishavane is already a drought prone area and farmers indicated that dry spells have become more frequent and devastating in recent years (Mutekwa 2009). Farmers in the area now experience extremely high temperatures and very low rainfall. Maddison (2006) also reported a similar trend of increasing temperatures in his study in 11 African countries. Another concern for smallholder farmers in Zvishavane is the changing onset of the rainfall season, which was confirmed by a study done by Mawere et al. (2013) in Zvishavane and Chivi. The study revealed that the climatic conditions that have prevailed for the past 5 years or so in agro ecological regions IV and V were very hot with dry conditions. There was rainfall below average (300 mm per annum) with poor distribution giving high chances of dry spells. In the past, the area used to receive three groups of rainfall, which are Bumharitsva (August), Hukurahundi (September), and Munakamwe (November) per year. Now they only receive one rainy season (munakamwe), which is also not predictable (Mawere et al. 2013). In addition, the farmers used to receive the last rains in March but this is no longer the case. The rains are no longer predictable as they can come very late and end early, something that has proved to be challenging to the farmers.

In the same study by Mawere et al. (2013), it was discovered that 96% of the respondents acknowledged their awareness of the changes in their local weather pattern and climate change. Many smallolder farmers in the study perceived the following as indicators: less and less rainfall leading to droughts, dwindling farming seasons, unpredictable weather patterns, high temperatures, decrease in livestock and crop production, low fruit production of wild fruit trees, and extinction of some area-specific species. Many smallholder farmers in Zvishavane have noted that although the coldest month of the year in Zimbabwe is June, it has been becoming less cold and the hottest month and October has become hotter than the previous years. It has also become apparent for most farmers that precipitation has been declining for years with more frequent drought occurrences. Drought occurrences have largely contributed to low yields and drastic effects on livestock production (Jiri et al. 2015). Farmers in Zvishavane have perceived inconsistent rainfall patterns and arid conditions in the area, which have led to the perennial flop of maize production (The Chronicle 2020). A survey done by International Crops Research Institute for the Semi-Arid Tropics (ICRISAT) noted that most farmers in Mberengwa and Zvishavane have been realizing very low yields from maize and other crops because of the effects of climate change (The Chronicle 2020).

Wetlands in Zvishavane have also declined over the years. Farmers in Zvishavane pointed out that their cattle used to drink water at wells and wetlands around the area but the wetlands are no longer there and they have to fetch water from distant places (Mawere et al. 2013). They also add that mountains in the area used to have wind storms signifying coming of heavy rains. This however no longer happens. Storms used to signify rainfall as well as animals such as ducks, baboons, rain bird-blue bird. Such animals are no longer existent as they have moved to other areas in search of food (Mawere et al. 2013).

Smallholder Solutions for Climate Change

Climate change has rendered Zimbabwe's region IV a non-maize producing zone due to climate shocks, a situation that promotes food insecurity (Mugiya and Hofisi 2017). This has left the majority of people in semi-arid areas food insecure, which calls for the need for adaptation. Given the reliance of most people in Zimbabwe on rain-fed agriculture, it becomes pertinent to identify solutions that can be implemented in order to deal with the insurmountable challenges posed by climate change. There are a couple of solutions that have been identified in some studies inclusive of conservation agriculture, irrigation, and gardening. A study by Mutasa (2011) showed that farmers in Buhera and Chikomba districts have incorporated certain strategies to address droughts that are prevalent in the area. The strategies are inclusive of early cropping, staggered cropping, dry planting, and planting crops with a short maturity life. This study recommends the production of small grains in semi-arid areas like Zvishavane. Small grain production will likely curtail the numerous challenges faced by farmers in Zvishavane given the advantages posed by small grains. Small grains have always been there and are not new as they were there before the introduction of maize; they are an indigenous African crop with more nutritional benefits than maize.

Several studies have looked at how small grains can lead to food security especially in semi-arid areas (Leuschner and Manthe 1996; FAO 2006; Mukarumbwa and Mushunje 2010). They highlight the numerous advantages associated with small grain production. Among such advantages are their nutritious value, drought resistance, and how they can be stored for longer periods as compared to maize. Leuschner and Manthe (1996) pointed out that sorghum and millet are some of the most important cereal crops for communal farmers in Natural Regions

1 V and V of Zimbabwe. Their study showed that in regions with low and erratic rainfall like Zimbabwe's Natural Regions 1 V and V, small grains have the potential of stabilizing household food security. In a study by Taylor (2003), sorghum and millet are vitally important cereals for the maintenance of food security in Africa. FAO (2006) also supports that small grains are the answer to chronic food shortages to rural communities who reside in semi-arid regions, especially of the sub Saharan region.

Characteristics and Impact of Small Grain Production on Climate Change Adaptation in Zvishavane District

As alluded to earlier, more than 475 million smallholder farmers across the globe cultivate less than 2 ha of land each (Lowder et al. 2016) on poor and marginal land with lack of access to technical or financial support that could help them invest in more climate-resilient agriculture (Morton 2007). These challenges are exacerbated by the effects of climate change, hence the need for appropriate adaptive and mitigation measure. Due to these challenges, smallholder farmers in semi-arid to arid areas continue to achieve very low yields. Growing small grains becomes an adaptation measure that smallholder farmers in areas prone to water and heat stress need to implement instead of maize, which is highly susceptible to water stress. Although maize is the preferred grain crop in Zvishavane and the rest of Zimbabwe, it is frequently written-off due to frequent droughts (Nciizah 2014) resulting in maize yields below 1 ton/ha. Consequently, small grain crops like millet and sorghum, which can better withstand drought conditions and offer more stable yields in the long term, are a better choice (Nciizah 2014; ICRISAT 2015). Small grains, particularly sorghum, adapt well to harsh climates and thus can grow in dry conditions due to their ability to tolerate heat, salt, and water stress, which makes them an ideal crop for semi-arid and arid areas (ICRISAT 2015).

There has been a conscious drive by the Government of Zimbabwe (GoZ) to urge farmers in arid and semi-arid to opt for small grain as a coping mechanism to climate change as evidenced by the increase in programs promoting sorghum production (GoZ 2014). The government of Zimbabwe has been working with international organizations such as FAO and ICRISAT to assist farmers in the country's marginal areas to focus more on the production of small grains such as sorghum and millet in order to counter poor yield and hence hunger. In Zimbabwe, small grains (sorghum, pearl, and finger millet) rank second after maize and consequently play a vital role in ensuring food security and nutrition (UNDP 2018b).

Droughts and dry spells have made smallholder farmers in Zvishavane to be expected to rely on irrigation schemes if they want to produce good maize yields. Unfortunately most of these farmers do not have the financial capacity to purchase the required equipment and the much needed fertilizers (Nciizah 2014). It is therefore vital for these farmers to concentrate more on small grain production compared to maize. Most of the farmers in the area are still concentrating on maize production with only a few household producing small grains. They still follow the norm of the country whereby maize is the mainstay crop even in drier areas where small grains can be produced economically and sustainably (Rukuni et al. 2006). However, this needs to change as maize production is no longer conducive for semi-arid areas that are constantly facing droughts and erratic rainfall. Recently, Zvishavane faced a drought in the 2019/2020 season leading to crop failure. In such circumstances, full adoption of small grain adoption becomes a necessity if the area is to avoid zero crop yields. This is mainly because small grains have characteristics and properties that allow them to thrive in areas like Zvishavane.

Sorghum and millet are important traditional cereal crops in Africa (Chisi Undated). Notably these cereals are indigenous to the African continent (Musara et al. 2019), making them well adapted to the African semi-arid areas (Taylor 2003). These small grains have become favored because of their good adaptation to hard environments and their good yield of production. The grains are drought resistant as compared to maize, making them to thrive in areas that are hot and have limited rainfall. Water requirements over the growing period average 400 mm for sorghum and 300–350 mm for millet as compared to 500 mm that is required for maize (Orr et al. 2016). Hence, they are genetically adapted to drylands that face little and irregular rainfall, drought, and high temperatures than other cereals like maize. This makes them to be able to give some yields in years of low rainfall, especially when grown in a multi-cropped system, whereas maize will be a complete failure (Muzerengi and Tirivangasi 2019).

Sorghum and millet also have deeper roots than maize and can withstand higher temperatures without damage to the crop (Orr et al. 2016). Bang and Sitango (2003) point out that small grains are generally the most drought-tolerant cereal grain crops as they require little input during growth. They bring more nutritional value to people's diet compared to cereals like maize. Dube (2008) posits that some of the advantages of small grains like sorghum and millet over maize include the following: a smaller amount of flour is needed to cook the main meal compared to maize, and a meal cooked from small grains satisfies hunger for a longer period and gives more energy (which is especially important for persons who do heavy manual labor like farmers).

Specifically, looking at finger millet, it can be observed that the crop has valuable properties that make it more beneficial to grow in the semi-arid of Zvishavane. The cereal is drought resistance in nature, has high nutritional content and has the ability to produce with few inputs. Finger millet is a nutritious crop providing proteins, carbohydrates, minerals, and amino acids, especially methionine, which is lacking in the diets of numerous poor people who live on starchy foods (Bhatt et al. 2003). This goes the same with sorghum, which remains important for sustainability of small-holder farmers' subsistence, social and economic livelihoods in semi-arid areas (Musara et al. 2019). Sorghum grain contains 11.3% protein, 3.3% fat, and 56–73% starch (National Research Council, 1996). It is relatively rich in iron, zinc, phosphorus, and B-complex vitamins. Tannins, found particularly in red-grained types, contain antioxidants that protect against cell damage, a major cause of diseases and aging (National Research Council, 1996). Thus, sorghum and millet have the capability of providing the nutritional value that can boost the immune system of people particularly poor families. They both have more health nutrients compared to maize, wheat, and rice and they improve the physical health of the elderly, the sick, children, and pregnant and nursing women. Another advantage of sorghum and finger millet maize is that they have a long shelf storage life as they can be stored for about 5 years compared to maize, which is only stored for about 2 years. Hence, they become conducive in Zvishavane, a place that is facing consistence drought and erratic rainfalls.

Although small grains are mostly suitable for Zvishavane, research conducted in the area shows that most smallholder farmers are still reluctant to fully adopt the grains. Mawere et al. (2013) in a study in Zvishavane stated that in as much as farmers reported that small grain crops performed much better than maize, most farmers still preferred growing maize. Similarly, Nciizah (2014) observed that an insignificant number of farmers were cultivating small grains in the area. Mugiya and Hofisi (2017) reported that while small grain crops prove to be a viable option to boost production in the background of climate change vulnerability, people in Zvishavane were not adopting such varieties enthusiastically, as adopting such varieties threatened their food preference. Hence, in Zvishavane, a significant number of households still prefer maize over small grain production. There are a number of barriers that have affected small grain production in Zvishavane. This has made the area to be consistently food insecure as smallholder farmers still grapple with inadequate yields.

Perceived Barriers to Small Grain Production in Zvishavane

Despite the advantages associated with small grains, small grain production remains low with only a few farmers producing these around Zimbabwe and specifically in Zvishavane. Studies done in Zimbabwe have shown that small grain production continue to decline (Gukurume 2013). The production of small grains has been very low in Zvishavane and the decline in small grain production is due to a number of barriers, which are central to this chapter. As indicated by smallholder famers in Zvishavane the major barriers to small holder production are labor constraints associated with small grain production, the challenge of birds (quelea birds), lack of inputs, and challenges associated with storage among other aspects. Studies have shown that small grain production is a good adaptive strategy, but is difficult to implement given the problems associated with the production of the crops. All barriers central to small grain production in Zvishavane are discussed below:

Lack of Inputs

UNDP (2018a) carried out a study on the barriers to small grain production. Their study showed that there are constraints in accessing inputs that limits the uptake of small grains. Overall, the results of the study showed that there is overreliance of farmers on untreated seed, evidencing inefficiency of the formal market. The study confirmed that treated and packed seed is not readily available in local markets or alternatively, if available, farmers do not afford the improved seed. This is the same case as what is transpiring in Zvishavane. There is limited access to small grain seeds, which reduces its production. In most cases, farmers end up using untreated seed, which largely contributes to poor yields.

While improved varieties of seeds have been released by NGOs in Zvishavane District, the seeds of these small grains are not readily available from the various seed producers (Nciizah 2014). Whether farmers in Zvishavane want to produce small grains or not, their choices are affected by limited seeds. Christian Care, a non-governmental organization (NGO) operating in Zvishavane, has distributed seeds to farmers but with limited stock, which has resulted in few farmers accessing it. In addition, given the limited seed stock of small grains and the easier accessibility of maize seeds most farmers end up preferring maize. Mugiya and Hofisi (2017) cited that in Zvishavane there is too much politicization of programs geared to boost productivity among small-scale farmers. Farmers in the area revealed that farm inputs (inclusive of seeds), which are meant to benefit smallholder farmers, are usually diverted for political initiatives impacting negatively on small grain produc-tion. In most cases farmers in Zvishavane end up planting sorghum seeds that they saved from their yield (Mugiya and Hofisi 2017). Most of these seeds are of very low quality, which impacts heavily yield.

Labor Intensiveness

The cultivation of small grains is extremely laborious, from land preparation, weeding, bird scaring to harvesting and grain processing. Most farmers in the area rely on manual production methods, which explains why labor intensiveness is a major concern. Nciizah (2014) noted that 95% of farmers in Zvishavane pointed to the labor intensiveness associated with small grain production as a very significant challenge. The study also showed how weeding is made more challenging by the similarities that exist between the weed "*Eleusine Indica*" and finger millet. The two are so similar that in most cases one might end up removing the crop and leaving the weed. There is also so much labor associated with harvesting the small grains given that the seeds are too small to handle and so much is done manually. Because the seeds are small it takes skill and much effort to mill finger millet (Nciizah 2014; UNDP 2018a). Farmers also mentioned the labor demands associated with pro-cessing of small grains as one of the limiting factors. Farmers in the study mentioned how processing of small grains is laborious as it involves threshing and winnowing, which is also done manually.

The Challenge of Quela Birds

Even though small grains are generally tolerant to diseases there is a huge challenge of damage by quela birds. Like in most studies (Nciizah 2014; UNDP 2018a), birds and animals are the major challenges in small grain production. According to

Nciizah (2019), the birds are attracted by the grain of the crops as the grains are exposed unlike those of maize, which is covered by leaves. Chasing birds is highly regarded as a strenuous activity that most farmers cannot embark on. Farmers in Zvishavane prefer producing maize as maize is prone to attacks by baboons and wild pigs, which are easier to scare away than birds which attack the crop in large numbers and are difficult to chase away (Nciizah 2014).

Low Yields

In Zimbabwe, production of maize continues to dominate in the country's semi-arid regions as compared to small grains (Sukume et al. 2000). The lower productivity associated with small grains makes them very unattractive to communal farmers in semi-arid areas. Small grains also require large farm size, which is out of the reach for most farmers in Zvishavane. Given the limited farm size in most places in Zvishavane, most farmers then prefer producing maize. A few of those who produce small grains have large land size and can also provide labor required to produce these. Some studies have also discovered that small grains do not yield much crop residue as compared to maize, resulting in most farmers preferring maize production. According to Mapfumo et al. (2005), maize production provides crop residue for livestock, which the livestock largely depend on for survival during winter.

Lack of Knowledge

Despite the evidently dismal performance of maize in the area as a result of the effects of climate change, most farmers have for years been disregarding and fiercely resisting advice from agricultural experts to plant small grains (The Chronicle 2020). Research has shown that people tend to be largely ill-informed about small grain production. Many farmers in Zvishavane ignored the calls to adopt small grain production and in many instances continued to produce maize despite the realization that they are likely to get very low yields due to drought (Nciizah 2014). This is largely attributed to taste preferences and the low yields associated with small grains. Due to unawareness and lack of prioritization of small grains, most farmers end up cultivating small grains in their worst part of arable land, which heavily impacts production yields.

Lack of Markets

It has been found out that when it comes to selling small grains, it has a very low market as people prefer other agricultural produce like maize. This is also confirmed by a study done by FAO (1996), which explains that as incomes rise, consumers tend to purchase wheat, rice, and, in some cases, maize, rather than traditional coarse grains. This has resulted in farmers viewing small grains as having lower earnings than other crops. Nciizah (2014) showed that one major disadvantage of small grains cited by Zvishavane farmers is the limited marketing opportunities. In the study farmers complained about the absence of a ready market for small grains. Farmers in Zvishavane do not rely on the Zimbabwe Grain Marketing Board (GMB) to buy small grains. According to Mugiya and Hofisi (2017), farmers complained that the GMB normally delays to pay farmers, a situation which further complicates their adaptation, as their purchasing power remains poor.

Low Extension Services and Government Support

Mukarumbwa and Mushunje (2010), show that small grain production has been negatively affected by inadequate government support. Research has shown that sorghum and millet are the most drought-tolerant cereal grain crops suitable in semi-arid regions (Taylor et al. 2006). The agricultural extension services department (AGRITEX) and the Department of Meteorological Services have extensively been incapacitated as a result of the country's economic and political challenges over the past decades that resulted in brain drain of skilled staff (Mutasa 2011). This has had a negative impact on farmers' production capacity and delivery of information about appropriate seeds and production technology. Mukarumbwa and Mushunje (2010) in their study indicated that inadequate government support to promote small grain production has led to low productivity of these small grains. Mugiya and Hofisi (2017) indicated that NGOs in Zvishavane are allegedly accused of imposing programs on small-scale farmers. The study also revealed that this also applies to government officials like extension officers. Extension officers believe that small-scale farmers are rigid and thus end up forcing farmers to adopt small grains. The farmers end up facing a challenge as their perceptions end up being ignored.

Food Preferences

A significant production constraint toward small grain production is food preferences. Most Zvishavane farmers significantly prefer maize due to taste preferences as compared to small grains. Mugiya and Hofisi (2017) reported that while small grain crops prove to be a viable option to boost production in Zvishavane in light of climate change, most farmers are not adopting small grains enthusiastically as it threatens their food preference. Such preferences affect the overall production of small grains. There remains a strong inclination to maize production as compared to small grain production. Maize is easily processed compared to small grains, which makes maize widely accepted. Jones (2011) pointed out food taste, media, and education as influential to people's food preference. Mugiya and Hofisi (2017) also indicated that children prefer maize to millet and sorghum. Food security in Zvishavane has been hindered due to farmers' rejection to produce small grains.

Interventions to Overcome the Barriers to Small Grains Adoption

Despite the huge potential of small grain crops to provide a viable adaptation mechanism to climate change, adoption among smallholder farmers remains disappointingly low due to the barriers addressed above. There is need for practical solutions from all stakeholders that will make the crop attractive for smallholder farmers in drought prone regions such as Zvishavane District. A comprehensive review by UNDP (2018b) on emerging solutions to small grains value chains in Zimbabwe identified the following interventions;

 (i) Availing improved varieties
 (ii) Improved processing methods and equipment
 (iii) Improved post-harvest management
 (iv) Improved access to markets for both inputs and outputs

These interventions are likely to make small grain crops more attractive for smallholder farmers. For instance, Musara and Musemwa (2020) showed that the allocation of more land toward improved sorghum varieties by smallholder farmers resulted in improved food diversity and food access. Moreover, improved varieties are likely to be more resilient to elevated temperature and low rainfall conditions due to climate change. However, it is necessary for stakeholders to increase funding for the development of improved small grains varieties, which currently lags behind of maize. Nevertheless, it is important that breeders, seed companies, seeds suppliers, processors, and retailers work closely with smallholder farmers to effectively understand and incorporate farmers' preferences as the end users.

It is also important to note that while small grains like sorghum perform comparatively better than maize under harsh conditions, unfavorable soil conditions, extremes in weather, pests, and poor management practices negatively affect yield. Consequently, it is important to complement improved varieties with sustainable cropping practices. One of the most widely recommended approaches is climate smart agriculture (CSA), which entails agricultural practices that concurrently increase productivity, strengthens resilience to climate change (adaptation), reduces/removes GHGs (mitigation), and contributes to achieving food security and development objectives (FAO 2010). Examples of such practices include conservation agriculture, legume intercropping, agroforestry, organic agriculture, and improved pest, water, and nutrient management. These practices significantly improve soil health, which ultimately improves crop productivity. Improved cropping practices have a potential to improve yields, which may in turn increase adoption of small grains. An increase in the adoption of the crop may in turn make seed production more attractive for established seed companies, who currently do not see the benefit of producing small grain seed due to low sales. However, for farmers to adopt these sustainable agricultural practices, there is a need to strengthen technical support services. Poor technology as well is generally one of the most prevalent challenges to agricultural production and uptake of new technology in smallholder farming areas.

Conclusions

The Chapter examined the role played by small grains in adapting to climate change in Zvishavane. Farmers in Zvishavane have perceived climate change as they have noticed changes in rainfall and temperature patterns in the past years. Such perceptions have been helpful in assisting many farmers to realize the need to adapt to climate change. The most appropriate way to adapt to climate change in the area as shown in this chapter is small grain production. Small grain production is the best strategy given that small grains are drought resistant and can withstand the hot temperatures in Zvishavane. However, despite this realization, an insignificant number of Zvishavane farmers are involved in small grain production. This is due to numerous barriers that have affected the smooth implementation of small grains. Such barriers are inclusive of the labor associated with small grain production, the challenge posed by the quelea birds, food preferences, low markets, and low extension services and government support. If nothing is done to mitigate these barriers, Zvishavane will continue to be a food insecure area. In this regard, the chapter highlighted possible interventions that can assist in helping farmers adopt small grains, which include the development of improved varieties, adoption of CSA practices, improved technical support, and access to markets. The labor intensiveness associated with small grain production requires technology and machinery that help in reducing manual labor. There is also need for policies promoting small grain production and processing to add value to small grains.

References

Aaron J (2012) A framework for the development of smallholder farmers through cooperative development. Directorate Co-operative and Enterprise Development Department: Agriculture, Forestry and Fisheries Republic Of South Africa. Available at: https://www.nda.agric.za/doaDev/sideMenu/cooperativeandenterprisedevelopment/docs/FRAMEWORK-%20OF%20SMALL%20FARMERS%20(2).pdf

Bang K, Sitango K (2003) Indigenous drought coping strategies and risk management in Papua New Guinea, Cgprt Monograph 43

Bhatt A, Singh V, Shrotria PK, Baskheti DC (2003) Coarse grains of Uttaranchal: ensuring sustainable food and nutritional security. Indian Farmer's Digest, Luchnow

Chisi (Undated) Sorghum and Millet Breeding In Southern Africa in Practice, Golden Valley Research Station, www.afripro.org.uk

Dube C (2008) The impact of Zimbabwe's Drought Policy On Sontala Rural Community In Matabeleland South Province, Msc Thesis, Department Of Geology, Geography And Environmental Studies, Stellenbosch University

FAO (1996) Rome declaration on world food security and world food summit plan of action. World food summit 13–17 November 1996, Rome

FAO (2006) Fertilizer use by crop in Zimbabwe. Rome, Italy

FAO (2007) Climate change and food security, in United Nations Joint Press Kit for Bali Climate Change Conference, December 3–14, 2007, Bali

FAO (2010) Climate-smart agriculture policies, Practices and Financing for Food Security, Adaptation and Mitigation. Rome, Italy. http://www.fao.org/docrep/013/i1881e/i1881e00.pdf

Gbetibouo GA (2009) Understanding farmers' perceptions and adaptations to climate change and variability: the case of the Limpopo Basin, South Africa IFPRI discussion paper. International Food Policy Research Institute, Washington, DC, forthcoming

Government of Zimbabwe (GoZ) (2014) Zimbabwe's climate change response strategy. Government of Zimbabwe, Ministry of Environment, Water and Climate, Harare

Gukurume S (2013) Climate change, variability and sustainable agriculture in Zimbabwe's rural communities. Rus J Agric Soc Econ Sci 14(2):89–100

HLPE (2013) Investing in smallholder agriculture for food security: a report by the high level panel of experts on food security and nutrition. HPLE report no. 6. Food and Agriculture Organization, Rome

Integrated Phase Classification Acute Food Insecurity Analysis Zimbabwe (IPC) (2019) Integrated food security phase classification: evidence and standards for better food security and nutrition decisions. FAO, Rome

Intergovernmental Panel on Climate Change (IPCC) (2001) Climate change 2001: impacts, adaptation vulnerability. Contribution of working group II to the third assessment report of the intergovernmental panel on climate change. Unep/Wmo, Geneva

Intergovernmental Panel on Climate Change (IPCC) (2007) Climate change: synthesis report. Contributions of working groups I, II and III to the fourth assessment report of the intergovernmental panel on climate change. IPCC, Geneva

International Crops Research Institute for the Semi-Arid Tropics (ICRISAT) (2015) "Is conservation agriculture an option for vulnerable households?" Briefing note 4, September. ICRISAT. Bulawayo, Zimbabwe

de Jager A, Onduru D, Van Wijk MS, Vllaming J, Gachini GN (2001) Assessing sustainability of low-external input farm management systems with the nutrient monitoring approach: A case study in Kenya. Agric Syst 69:99–118

Jerie S, Ndabaningi T (2011) The impact of rainfall variability on rainfed tobacco in Manicaland province of Zimbabwe. J Sustain Dev Afr 13(1):241–250

Jiri O, Mafongoya P, Chivenge P (2015) Smallholder farmer perceptions on climate change and variability: a predisposition for their subsequent adaptation strategies. J Earth Sci Clim Change 6:277. https://doi.org/10.4172/2157-7617.1000277

Jones S (2011). How does a food systems approach elucidate the food insecurity of Inuit in Canada? Global Environmental Change and Food Systems, GECAFS Working Paper7. University of Oxford

Leuschner K Manthe CS (1996) Drought tolerant crops for Southern Africa. In: Proceedings of the SADC/ICRISAT regional sorghum and pearl millet workshop, 25–29 July 1994. Gaborone, Botswana

Louw A (2013) Sustainable policy support for smallholder agriculture in South Africa: key issues and options for consideration. In: Greenburg S (ed) Smallholder and agro-food value chains in South Africa. PLAAS, Bellville

Lowder SK, Skoet J, Raney T (2016) The number, size, and distribution of farms, smallholder farms, and family farms worldwide. World Dev 87:16–29

Maddison D (2006) The perception of and adaptation to climate change in Africa. CEEPA discussion paper no. 10. Centre for Environmental Economics and Policy in Africa, University of Pretoria

Mapfumo P, Mtambanengwe F, Giller KE, Mpepereki S (2005) Tapping indigenous herbaceous legumes for soil fertility management by resource-poor farmers in Zimbabwe. J Agric Ecosyst Environ 109:221–233. [Online]. Available from: http://www.sciencedirect.com

Mawere M Madziwa BF Mabeza CM (2013) Climate change and adaptation in third world Africa: a quest for increased food security in semi-arid Zimbabwe. Int J Human Soc Stud. Issn 2321–9203

Morton J (2007) The impact of climate change on smallholder and subsistence agriculture. PNAS 104:19680–19685

Mugabe FT Hodnett M Senzanje A (2007) Comparative hydrological behaviour of two small catchments in semi-arid Zimbabwe. J Arid Environ. https://doi.org/10.1016/J.Jaridenv.2006.11.016

Mugiya D, Hofisi C (2017) Climate change adaptation challenges confronting small scale farmers. Environ Econ 8(1):57–65. https://doi.org/10.21511/ee.08(1).2017.06

Mukarumbwa P, Mushunje A (2010) Potential of sorghum and finger millet to enhance household food security in Zimbabwe's semi-arid regions: a review, contributed paper presented at the joint

3rd African Association of Agricultural Economists (AAAE) and 48th Agricultural Economists Association of South Africa (AEASA) Conference, Cape Town, South Africa, September 19–23, 2010

Musara JP, Musemwa L, Mutenje M, Mushunje A, Pfukwa C (2019) Determinants of sorghum adoption and land allocation intensity in the smallholder sector of semi-arid Zimbabwe. Span J Agric Res 17(1):E0105. https://doi.org/10.5424/Sjar/2019171-13115

Musara JP, Musemwa L (2020) Impacts of improved sorghum varieties intensification on household welfare in the mid-Zambezi Valley of Zimbabwe. Agrekon 59(2):254–267

Mutasa M (2011) Climate change vulnerability and adaptation in failing states: Zimbabwe's Drought struggle: Paper prepared for the Initiative on Climate Adaptation Research and Understanding through the Social Sciences (ICARUS-2) meeting at the University of Michigan (5–8 May 2011) themed, "Vulnerability and Adaptation: Marginal Peoples and Environments." Although there are several climate change-induced impacts, this paper focuses on drought, which poses a huge challenge in rural rain-fed agro-economies

Mutekwa VT (2009) Climate change impacts and adaptation in the agricultural sector: the case of smallholder farmers in Zimbabwe. J Sustain Dev Afr 11(2):237–256

Muzerengi T, Tirivangasi HM (2019) Small grain production as an adaptive strategy to climate change in Mangwe District, Matabeleland South in Zimbabwe. Jàmbá J Disaster Risk Stud 11(1):A652. https://doi.org/10.4102/Jamba.V11i1.652

Nciizah T (2014) The contribution of small grain production to food security in drought prone areas: the case of Zvishavane (2000–2014) A thesis Submitted in Fulfilment of the Requirements for the Degree of Master of Arts in Development Studies at Midlands State University

Nciizah E. (2019). Understanding climate change and rural livelihoods in Zimbbabwe: adaptation by communal farmers in Ngundu, Chivi District, A thesis Submitted in Fulfilment of the Requirements for the Degree of Doctor Philosophy of Rhodes University

Oakland Institute and The Alliance For Food Sovereignty In Africa (AFSA) (Undated). www.oaklandinstitute.org, www.afsaafrica.org

OCHA (2012) Midlands province – natural farming regions. http://ochaonline.un.org/zimbabwe

Okonya JS, Syndikus K, Kroschel J (2013) Farmers' perception of and coping strategies to climate change: evidence from six agro-ecological zones of Uganda. Received: OnlinePublished: July15, 2013. https://doi.org/10.5539/jas.v5n8p25

Orr A, Mwema C, Gierend A, Nedumaran S (2016) Sorghum and millets in Eastern and Southern Africa. Facts, trends and outlook. Working paper series no. 62. ICRISAT research program, markets, institutions and policies. Patancheru 502 324. International Crops Research Institute for the Semi-Arid Tropics, Telangana. 76 Pp

Rukuni M,Tawonezvi P, Eicher C, Munyuki-Hungwe M, Matondi P (2006) Zimbabwe's agricultural revolution revisited, University of Zimbabwe Publications. Harare, Zimbabwe

Sharma KK, Crouch JH, Seetharama N, Hash CT (2002) Applications of biotechnology for crop improvement: prospects and constraints. Vitro Cell Dev Biol—Plant 163:381–395

Simba F, Chikodzi D, Murwendo T (2012) Climate change scenarios, perceptions and crop production: a case study of semi-arid Masvingo province in Zimbabwe. J Earth Sci Clim Change 3:124. https://doi.org/10.4172/2157-7617.1000124

Sukume C, Makudze E, Mabeza-Chimedza R, Zitsanza N (2000) Comparative economic advantange of crop production in Zimbabwe. Technical paper no. 99 Department of Agricultural Economics and Extension. University of Zimbabwe, Harare

Taylor JRN (2003). Overview importance of sorghum in Africa [Online]. Available from: http://www.sciencedirect.com/science?

Taylor JR, Schober TJ, Bean S (2006). Novel and non-food uses for sorghum and millets. J Cereal Sci 44:252–271

The Chronicle (2020) Promotion of small grains pays dividends for farmers: Chronicle.co.zw

UNDP (2018a) Barrier analysis of small grains value chain in Zimbabwe, Technical Notes Series No. 3

UNDP (2018b) Emerging solutions in small grains value chain in Zimbabwe, Technical Notes Series No. 4

Zimbabwe Agricultural Sector Survey (2019) Edited by Mutenga T, The State of Zimbabwe's Agricultural Sector Survey 2019, Harare

Zimbabwe Vulnerability Assessment Committee (ZimVAC) Market Asseessment Report (2015) Food and Nutrition Council at SIRDC, Hatcliffe, Harare, Zimbabwe

Zimbabwe Vulnerabilty Assessment Committee (ZimVAC), Rural Livelihoods Assessment (2016) Food and Nutrition Council (FNC) at SIRDC, Hatcliffe, Harare, Zimbabwe

2

Use and Impact of Artificial Intelligence on Climate Change Adaptation in Africa

Isaac Rutenberg, Arthur Gwagwa and Melissa Omino

Contents

Introduction
Scoping AI in Afric
    Case Studies of General Applications of AI in Africa
Review of African AI Focusing on Climate Change Issues
    Geospatial Technologies
    The Private Sector, and Converging Exponential Technologies
    University Activity
    The Potential of A
Challenges and Future Applications
    Scope of the Problem
    Ethical Issues of Predicting Climate Change Impacts
    Data Inadequacies
Conclusions
References

Abstract

Although Climate Change is a global phenomenon, the impact in Africa is anticipated to be greater than in many other parts of the world. This expectation is supported by many factors, including the relatively low shock tolerance of many African countries and the relatively high percentage of African workers engaged in the agricultural sector. High-income countries are increasingly turning their focus to climate change adaptation, and Artificial

I. Rutenberg (✉) · A. Gwagwa · M. Omino
CIPIT, Strathmore University, Nairobi, Kenya

e-mail: irutenberg@strathmore.edu

Intelligence (AI) is a critical tool in those efforts. Algorithms using AI are making better predictions on the short- and long-term effects of climate change, including predictions related to weather patterns, floods and droughts, and human migration patterns. It is not clear, however, that Africa is (or will be) maximally benefitting from those AI tools, particularly since they are largely developed by highly developed countries using data sets that are specific to those same countries. It is therefore important to characterize the efforts underway to use AI in a way that specifically benefits Africa in climate change adaptation. These efforts include projects undertaken physically in Africa as well as those that have Africa as their focus. In exploring AI projects in or about Africa, this chapter also looks at the sufficiency of such efforts and the variety of approaches taken by researchers working with AI to address climate change in Africa.

Keywords

Africa · Climate change · Artificial intelligence · Algorithms · Data · Adaptation · Migration

Introduction

The deepest roots of climate change begin with the second industrial revolution and the widespread adoption of fossil fuel-based machinery. As the world enters the fourth industrial revolution, the adoption of advanced technologies such as artificial intelligence (AI) introduces complex challenges and opportunities for the now-inevitable and as-yet-undetermined issues of climate change. This chapter explores those challenges and opportunities, and whether or to what extent the fourth industrial revolution will enhance Africa's ability to cope with climate change.

Technologies associated with the fourth industrial revolution (4IR) include blockchain, the Internet of things (IoT), artificial intelligence, cloud computing, quantum computing, advanced wireless communications, and 3D printing, among others. Although these technologies are, at times and in various ways, interrelated, this chapter focuses mainly on AI and the impact that it will have on Africa's ability to cope with climate change. This focus is not, however, meant to imply that other 4IR technologies will not be important. Indeed, some technologies such as the widespread connectivity of sensors that accompany the development of IoT will likely have substantial positive impact on the ability of African nations to gather useful data. An example of this is Upepo Technologies, described below, that is using IoT devices to monitor water distribution points throughout Kenya. Other technologies may have mixed or negative impact: cryptocurrencies based on blockchain technology have so far increased the amount of energy used worldwide by computers, and advanced communications technologies are projected to account for 20% of global energy demand by 2025 (The Guardian 2017). Any increase in energy demand has a negative effect on climate change. Surely these technologies will continue to evolve, and Africans' ability to predict their ultimate impacts is limited.

AI has probably been postulated for longer than any other 4IR technology, but has remained impractical until the recent decade. Although a variety of forms of AI are known and being pursued in various contexts and by various private- and public-sector stakeholders, this chapter is largely confined to discussing AI as a tool for processing large amounts of data and improving predictions from those data.

Predictive algorithms using AI are available to entrepreneurs using two main pathways. The first is by using and adapting widely available open source algorithms. Many of the global technology companies have released suites of AI tools on open source licenses, including Google, Microsoft, and IBM. The alternative pathway is to develop, test, and train novel algorithms. Examples of companies following each of these pathways, as well as combinations thereof, can be readily identified, particularly (but not exclusively) in the high technology hubs of Lagos, Nairobi, and Johannesburg.

Developing the algorithms is, however, only one part of a larger process of integrating AI into products and processes. Effective predictive algorithms using AI rely on three main components: fast computers, large datasets, and abundant human labor (The Economist 2020). The need for fast computers is obvious, and suitable computing power is relatively abundantly available, even if Africa is perpetually and conspicuously absent from international lists of supercomputers (TOP500.org 2019). Human labor is needed in order to label data, as AI algorithms require accurately labelled data in order to learn and increase the accuracy of predictions. Africa has an abundance of labor, including large numbers of unemployed or underemployed youth that have sufficient computer skills, and is therefore ideally positioned for performing this task. For example, Samasource is a data annotation provider with a major presence in Kenya and Uganda. Using a hybrid structure involving both for-profit and nonprofit entities, Samasource provides data labelling services for Fortune 500 companies employing AI systems, while simultaneously seeking to reduce poverty through job creation.

This leaves three factors that are particularly interesting in the African context: the availability of sufficiently large datasets, the suitability of the algorithms themselves; and the creativity of developers, entrepreneurs, and others in applying AI in the creation of new products, services, and solutions. The three factors of interest are interrelated, and it is difficult to assess any one factor in isolation. AI is a multipurpose tool, but the output of any given AI product is as much or more dependent upon the dataset as on the specific AI algorithm. Regardless of the algorithms, the availability of high quality, locally relevant datasets is likely to be extremely important for AI to be of any use to the fight against climate change. Furthermore, since the bulk of AI development has so far been from outside the African continent, the creativity and effort of developers and their collaborators are also critical to the contribution of AI. Accordingly, the second (datasets) and third (applications) factors are the subject of this chapter.

Part I of this chapter introduced the topic and framed the research question. Part II broadly and briefly review the state of AI research and development in Africa, as well as global AI research that is relevant or targeted at Africa. Part III narrows the focus and explore African AI products and research that are specific to the issues of Climate Change. Such issues include weather prediction, agriculture, floods and drought, and human migration. The chapter concludes in Part IV, with a look at particular challenges and potential future applications of AI in the area of climate change.

Scoping AI in Africa

It is easy to get the impression that there is little or no activity in AI research and development in Africa. In report after report, African countries are either characterized as poorly performing or are complete absent from the analysis. African governments score very low in their perceived readiness to "take advantage of the benefits of AI in their operations and delivery of public services" (Oxford Insights International Development Research Centre 2019). In that global index, no African government ranked within the top 50 governments. In a broader index covering 54 countries over seven factors (Talent, Infrastructure, Operating Environment, Research, Development, Government Strategy and Commercial Ventures), all six African countries that were part of the study ranked in the bottom quartile (Tortoise Intelligence 2019). In an even broader analysis using different datasets to "comprehensively assess the state of AI R&D activities around the world," the African region was ranked at or near the bottom of nearly every measure and ranking (Perrault et al. 2019). This included measurements such as AI papers published and cited, AI-focused patents, conferences, and technical performance. The picture painted by these global indices is one of virtually no activity and a gloomy outlook for future activity.

As is often the case, however, the global indices are not necessarily good measures of activity on the African continent. Activities by African governments and certain other entities are often not reported or described online. Activities in the private sector may be minimal when compared to activity levels in developed countries, but the comparison is not particularly useful. As shown by the examples provided below, the R&D that is carried out is meaningful and a significant part of the overall research environment on the continent.

African-hosted conferences on AI showcase African-led and Africa-focused AI R&D and demonstrate the breadth of such work. The Deep Learning Indaba, which began in 2017, is such an example; devoted entirely to African AI R&D, it has doubled attendance each of the following 2 years. The yearly event has created great interest, and in 2018, 13 mini-conferences (referred to as "IndabaX" conferences) were held in as many countries in Africa. Such conferences are important for the normal reasons, but also because African researchers have particular difficulties attending AI conferences in other parts of the world. African nationals are routinely denied visas to attend international conferences (Knight 2019). The challenges faced by African nationals is significant enough that the International Conference on Learning Representations, a major AI conference, elected to hold the 2020 conference in Africa. This decision was intended to enable greater African participation (Johnson 2018).

Meanwhile, global technology companies are turning some of the focus of their AI R&D on Africa. Google opened an AI development laboratory in Ghana in 2019. Microsoft published a whitepaper focusing on the importance of AI research in Africa (Microsoft Access Partnership University of Pretoria 2018). A significant body of technological developments have been produced by IBM researchers at the IBM laboratory in Nairobi (Weldemariam et al. 2020). Motivations for global technology companies to work in Africa include, among others, the reduction of bias by increasing diversity in the researchers and the context in which they work (Adeoye 2019).

Case Studies of General Applications of AI in Africa

Academic and private sector AI researchers in Africa seek to apply AI to a wide variety of topics, some of which are directly relevant to climate change, and others are relevant insomuch as they advance the general state of AI. A particularly important topic is bioinformatics and genomics (Diallo et al. 2019). Significant biodiversity exists on the African continent, and much of it is understudied relative to other regions (Campbell and Tishkoff 2008). The application of AI algorithms, with its ability to analyze large datasets, is particularly appropriate in the area of genomics. Even where the work is not specifically addressing issues directly related to climate change, a better understanding of bioinformatics and genomics will support biodiversity and our ability to encourage genetic resistance to environmental shocks.

Another topic of interest to African AI researchers is farming. The importance and vulnerability of farming on the African continent cannot be overstated. More than 50% of the entire African workforce is employed in agriculture, yet most of that activity is at the subsistence level (Alliance for a Green Revolution in Africa (AGRA) 2014). Approximately 80% of all farms in Sub-Saharan Africa are smallholder farms where 175 million people are directly employed. Modern farming methods such as large-scale mechanized harvesting, irrigation, and crop rotation are less common, meaning that African agriculture is more susceptible to weather and labor shocks. Traditional African crops are under threat from global seed companies seeking to dominate the African market (Scoones and Thompson 2011). Deforestation and overgrazing threaten African soil, but productive farmland is targeted by foreign governments and companies seeking new resources to enhance food security (The World Bank 2014).

It is anticipated that predictive analytics, aided by machine learning, will significantly improve overall output by the African farming sector (Alliance for a Green Revolution in Africa (AGRA) 2014). The cornerstone of successful AI is sufficient data, and in this area, Africa is relatively well positioned. Datasets pertaining to agriculture are available from a variety of sources, as the subject of African agriculture has been of intense interest to both local and international stakeholders for many decades. These datasets include actual measurement data as well as predictive or extrapolative datasets, as well as combinations thereof (African Soil Information Service). Improving the quantity and quality of data, and encouraging wider and more effective use of Big Data, is the focus of research by Patrick MacSharry at the Carnegie Mellon University campus in Kigali, Rwanda. Numerous other academic researchers are also contributing to the push toward effective incorporation of data science in African agriculture.

Successful agriculture requires more than land, seeds, and suitable weather. Crop diseases and infestations of plant pests are significant threats to global agriculture and are particularly dangerous to agriculture in Africa. For example, invasions of the desert locust can last for up to a decade, and can cause widespread food shortages of catastrophic proportions (Lecoq 2003). Similarly, strains of wheat rust such as ug99, which originated in Uganda, are considered a global threat as well as a threat to African agriculture (Aydoğdu and Boyraz 2012). An ability to diagnose crop diseases is therefore critical, and local and global technology industries have taken notice. Mobile phone applications that allow farmers to diagnose crop disease or identify the presence of crop pests are proliferating (Nzouankeu 2019). Considering that many plant diseases and crop pests are highly geographically specific, local development of technology solutions to these problems is a prerequisite to their success, and the African technology industry is beginning to address this need.

Looking beyond agriculture, financial institutions in Africa are not often at the forefront of widespread introduction and adoption of technology. Notwithstanding the success of mobile banking (which more properly attributes success to the fact that it originated from the telecom industry rather than the banking industry), traditional financial institutions have been slow to adopt technologies such as online banking. This is not necessarily only due to resistance within the African banking sector, but is also due to the relatively low level of access to technology by banking customers. Nevertheless, nearly every industry sector in Africa is affected by the availability (or absence) of credit and the presence (or absence) of financial institutions. Financial technologies, referred to as "fintech," are a particularly active area of the incorporation of AI into products and services. A common use of AI in fintech is in the determination of creditworthiness. Although not discussed here, it is worth noting that such applications have been criticized for a variety of reasons, ranging from data protection issues to exacerbation of inequality (Johnson et al. 2019). Traditional and nontraditional banking institutions can make use of larger datasets from a broader list of sources in order to determine the risk of extending credit to individual or institutional customers. This is particularly useful in a region where formal credit histories are relatively rare or based on very limited data. Another fintech product that is gaining traction with African financial institutions is the use of AI to evaluate customer behavior. Behavior analysis can be used to improve predictions of suitable financial products, thus allowing institutions to focus their marketing efforts and improve efficiencies.

Raising capital to support product development and market expansion is a necessary component of the modern technology industry. There is significant evidence that success at attracting investor attention in Africa is due to a variety of factors, including some factors that are not related to the quality or appropriateness of the technology (Peacock and Mungai 2019). Nevertheless, some African startups have raised significant capital for the application of AI to agriculture, fintech, and mobility solutions, among other uses of AI. In South Africa, Aerobotics is a company that is developing a digital platform for using AI to interpret data gathered by drones and satellites. The data interpretation algorithms are targeted at detecting plant pests and diseases. Apollo Agriculture is a Kenyan company using AI to interpret satellite data, soil data, farmer behavior, and crop yield models. The data interpretation algorithms allow the company to offer farmers customized financing as well as customized seed and fertilizer packages (Nanalyze 2019).

In view of the above examples and others not mentioned, there is clearly activity in AI R&D in Africa. It may be that the activity is minimal when compared with that of developed countries, but there is a very high level of interest in AI technologies by governments, academia, and the private sector, and activity levels will surely increase with time.

Having reviewed the general state of AI R&D in Africa, and having established that there is ongoing interest and activity in applying AI to a variety of fields, this chapter now turns to the specific application of AI to climate change.

Review of African AI Focusing on Climate Change Issues

By 2050, it is widely expected that hundreds of millions of people in developing countries will have left their homes as a result of climate change – a mass displacement that will make already-precarious populations more vulnerable and impose heavy burdens on the communities that absorb them. Unfortunately, the world has barely begun to prepare for this impending crisis.

According to the World Politics Review, those displaced by climate change are neither true refugees nor traditional migrants, and thus occupy an ambiguous position under international law. Consequently, the world needs to agree on how to classify environmental migrants, as well as what their rights are. It also needs to strengthen its capacity to manage these mass migrations, without weakening existing international regimes for refugees and migrants (Patrick 2020).

Similarly, international organizations, such as the United Nations, as well as industry and social society are exploring technological solutions including the development of tools and technologies that leverage data sources from radio content, social media, mobile phones and satellite imagery, and technology toolkits. These toolkits can enhance decision-making by providing real-time situational awareness for project and policy implementation. The trend is moving towards a converging of exponential technologies (AI, robotics, drones, sensors, and networks). Although there is growing criticism of this trend toward 'technologizing of care', which can conflict with the centrality of the humanitarian principles, these technologies, in particular AI, have the potential to help Africa to be better prepared to thwart the effects of climate change, both through mitigation and adaptation.

In light of this, this section examines the use of AI that is being (or could be) used in Africa for any aspect of climate change such as weather prediction, agriculture, human migration, floods and droughts. These projects could be undertaken physically in Africa as well as in those countries that have Africa as their focus, and those that can be adapted to Africa. It assumes a broader definition of AI which includes the use of "software (and possibly also hardware) systems designed by humans that, given a complex goal, act in the physical or digital dimension by perceiving their environment through data acquisition, interpreting the collected" (European

Commission High-Level Expert Group on Artificial Intelligence 2018). It examines how a combination of technologies work together, including geospatial technologies, converging technologies such as robotic responders, swarm technology, and aerial drones which may also rely on geospatial data. Further, it also examines AI-driven platforms that crowdsource first-hand experiential data from those on the ground, and how these technologies could potentially transform disaster relief methods in Africa. Finally, it examines the issue of mobile connectivity and wireless networking trends and how improving these can give a newfound narrative power to those most in need. Although some of the projects examined in this section do not directly address climate change, per se, they may address other environmental concerns that are relevant to climate change.

Geospatial Technologies

Africa has seen a steady increase in multi-stakeholder geospatial initiatives in recent years and these include those sponsored by the UN agencies, Public Private Partnerships, and initiatives led by international organizations and industry, especially global technology companies such as Google and Vodafone.

UN Agencies

The UN, in particular its data science arm, the Global Pulse, has been working to support African governments to achieve the U.N. 2030 Agenda for Sustainable Development. The Global Pulse, working from its Kampala Lab in Uganda, has led the work to develop numerous toolkits that consolidate it as an important technical arm of the UN network (Pulse Lab Kampala 2018). These pieces of software are a key in informing the SDGs through big data, data science and artificial intelligence because they aggregate, anonymize, combine, analyze, and visualize data. During 2016 and 2017 the Lab has both created brand-new toolkits and adapted previously developed ones for new projects. This has been part of a bigger project under which, between 2016 and 2017, the Pulse Lab Kampala worked with various UN agencies and development partners in Uganda and the region to test, explore and develop 17 innovation projects. The Lab also furthered the development of tools and technologies that leverage data sources from radio content, social media, mobile phones and satellite imagery, and created technology toolkits. These toolkits can enhance decision-making by providing real-time situational awareness for project and policy implementation (Pulse Lab Kampala 2018).

The Pulse's projects sit in the current global policy shift which emphasizes technological tools as opposed to their application to improving knowledge about climate change. This is even reflected in the secondary school geography curriculum which has seen an introduction of GSI in some African countries (Cox et al. 2014). This shift is borne from the premise that geospatial technologies, synergetic applications of remote sensing, and geographical information systems, offer versatile cross-scale tools to study long term climate changes, and their impacts on social- and ecological systems.

According to the UN World Data Forum, the U.N. 2030 Agenda for Sustainable Development requires countries to be chiefly responsible for collecting information and monitoring progress towards achieving economic, social, and environmental sustainability (United Nations World Data Forum 2018). For instance, the UN World Data Forum has held sessions to enable the sharing of sound Earth Observation (EO) methods, tools, and technologies, as well as national use cases of effective integration of EO and geospatial data products with traditional and other relevant information sources in support of the Sustainable Development Goals (SDGs), targets, and indicators, and for informed decision-making. Some of the technologies discussed at such forms include the African Regional Data Cube, aimed to enable greater use of EO and geospatial data for sustainable development (Anderson et al. 2017).

To strengthen its work at policy level in this area, in 2019, the UN Pulse has been engaging African AI experts to lead efforts to draft a Code of Ethics for the use of AI-supported systems in humanitarian and development contexts (International Telecommunication Union 2019). Once developed, the Code may be used as guidance for the work at Global Pulse and at other UN organizations deploying AI for social good and to ensure the deployment of such technologies is both ethical and human rights respecting.

Public-Private Partnerships

Similarly, the private sector has been working with governments on data science projects that leverage data for AI applications in humanitarian contexts. For instance, the Vodafone Foundation pioneered a program in Ghana to use aggregated anonymized data to help the government track and control epidemics to prevent widespread outbreaks (Vodafone 2018). The program, dubbed as one of the first of its kind globally, tracks and analyzes real-time trends in population movement. The program demonstrates the use of big data in situations directly relevant to climate change. The Ghana initiative is a good example of a multistakeholder approach to technology deployment for humanitarian ends as it is not only supported by Vodafone Ghana but also the Flowminder Foundation, an NGO that provides insights, tools and capacity-strengthening to governments as well as international agencies and NGOs (GhanaWeb 2019).

The South Africa company Aerobotics, mentioned above, operates a public private partnership that utilizes aerial imagery and machine learning algorithms to solve specific problems across some industries including insurance and agriculture. In May 2019, Aerobotics signed an agreement with Agri SA to offer free service for all South African farmers (Lukhanyu 2019). Drones are used to track tree health and size, using multispectral, high resolution imagery. The project also enables farmers to identify areas needing attention from historical satellite health data, and inspect these in the field using a mobile app. The Aerobotics project is supported by the South African Department of Environmental Affairs (DEA) which works with the Committee on Spatial Information (CSI) and the broader GIS community to define the data architecture, systems, standards, policies and processes for a fully integrated and effective spatial data infrastructure for the country. The Environmental Geographical Information Systems (E-GIS) webpage provides access to baseline environmental geospatial data, map services, printable maps and relevant documents to users of geospatial technology, government, and the public (Department of Environmental Affairs South Africa 2016).

Another example, also out of South Africa, applies machine learning to the issue of air quality prediction (Chiwewe and Ditsela 2016). Stemming from the Green Horizons initiative, IBM researchers partnered with Chinese government researchers for the purpose of building air quality prediction software. In Johannesburg, the work is to adapt the air quality prediction software to the local context. The Green Horizon's system harnesses historical and real time data about weather, air quality, topography, and traffic reports to build predictions about air quality. The project's South Africa lead, Tapiwa Chiwewe, says that the task is to tweak the software to local particularities. For instance, Johannesburg does not have the dense network of air quality monitoring stations (eight stations compared to 35). Chiwewe and the team of researchers sought to 'teach' the software to work with more sparse data and to use intermediate fixes to make up for the lack of data.

The Private Sector, and Converging Exponential Technologies

International Organizations & Global Technology Companies

Globally, converging exponential technologies (AI, robotics, drones, sensors, networks) are transforming the future of disaster relief (Diamandis 2019). African stakeholders have been experimenting with these technologies in a variety of contexts. These efforts have been led by international organizations such as Omdena and Element AI, working with local African NGOs like R365 and the Nigerian NGO Renewable Africa (Adewumi 2020). Academic institutions and global technology companies such as Google also play a part in this work, which span the R&D process as well as prototyping and implementation. For example, the Canadian based Element AI has African-focused projects that support the use of robots for humanitarian purposes. Their intention is to develop human-machine collaborations that build up a trusted relationship with AI products and services already available. Two further examples are illustrative. Atlas AI, a Silicon Valley public benefit corporation, has teamed with the Alliance for a Green Revolution in Africa (AGRA), to apply predictive analytics and machine learning to help process numerous datasets in an effort to improve smallholding farming output (AGRA 2019). Pennsylvania State University developed, deployed, and continue to upgrade PlantVillage, a mobile application that uses an AI tool to diagnose crop diseases in Africa (Penn State 2019).

Global initiatives inspired by Kenya's Ushahidi are emerging that leverage AI, crowd sourced intelligence, and cutting-edge visualizations to optimize crisis response (Starbird 2012). Such projects include One Concern (2020), which employs AI in analytical disaster assessment and damage estimates. Crowdsourced intelligence (which includes predictive crisis mapping and AI-powered responses) is used in response to both natural disasters and humanitarian disasters. An open-source crisis-mapping software developed by Ushahidi is used for real-time mining of social media, news articles, and geo-tagged, time-stamped data from countless sources (Meier 2012). As mobile connectivity and abundant sensors converge with AI-mined crowd intelligence, real-time awareness will only multiply in speed and scale (Diamandis 2019).

Other organizations are using similar crowdsourcing technologies to address different challenges, but such technologies are also helpful in understanding agricultural and other environment concerns. IBM's Hello Tractor is an open source mobile platform that enables farmers to access tractor services on demand (Assefa 2018). By using technology integrated from partners like the IoT companies Aeris and CalAmp, the platform can tell when a tractor is turned on and how far it travels. By using the platform, over the next 5 years, through a public-private partnership, John Deere plans to deploy 10,000 tractors in Nigeria, selling them to contractors who then rent them out to small farmers (Peters 2018). Considering that climate change is expected to increase uncertainty in the long-term viability of agricultural land, the availability of tractors for rent will be critical as a means for improving flexibility of farmers.

Another organization using AI-based crowdsourcing solutions to climate-related or climate-relevant challenges is Omdena, which sources ideas to respond to local challenges. Several of Omdena's projects are worth discussion. Under one of its challenges, 34 collaborators working together with the UN Refugee Agency (UNHCR) built several AI and machine learning solutions to predict forced displacement, violent conflicts, and climate change effects in Somalia (Omdena 2019). Their community of AI experts and data scientists have developed several solutions to predict climate change and forced displacement in Somalia, where millions of people are forced to leave their current area of residence due to natural and man-made disasters such as droughts, floods, and violent conflicts. This is a holistic project under which Omdena's challenge partner, UNHCR, provides assistance and protection for those who are forcibly displaced inside of Somalia. The findings will help UNHCR improve speed and efficiency of responses to such disruptions.

In a second project, Omdena's collaborators analyze conflict data to build a hot zone representation, which predicts the most dangerous locations and the highest fatalities (Omdena 2019). The machine learning model can help to optimize the allocation of utility personnel to handle incidents. A promising application of this technology is to leverage satellite images to assess the environmental impact of forced displacement and conflict by comparing the weekly Vegetation Health Index with human displacement data.

Omdena's projects are also focused on increasing the adoption of renewable energy, an important component of climate change mitigation. In Nigeria, Omdena's AI community built an interactive map showing the top Nigerian regions for solar power instalments (Adewumi 2020). The solutions will provide helpful insights for government and policy makers to make decisions on where to allocate resources in the most effective way. In a country where more than 100 million people lack stable access to electricity, renewable energy must be a major part of any environmentally friendly solution. The Omdena community generated a variety of outputs, including a grid coverage analysis and machine-learning-driven heatmaps to identify sites that are most suitable for solar panel installation. Along with an interactive map listing the top Nigerian regions in terms of demand for electricity, such tools are helpful for those seeking to survey and validate locations before installing solar panels. This will enable data-driven investments and policy-making and potentially impact the lives of many people in Nigeria.

A particularly forward-looking initiative is the Microsoft AI for Earth grant program. One of the recipients of the grant is Upepo Technology, a Kenyan company that plans to use the grant in a water monitoring project. The company is deploying a large network of IoT devices, and employing AI algorithms to analyze the data from sensors monitoring reservoirs, boreholes, water kiosks, individual taps, and other water points. Considering the substantial impact that climate change has on issues pertaining to water – particularly by changing the patterns of precipitation – an enhanced ability to monitor water usage, wastage, and storage will greatly benefit the ability to deal with climate change impacts.

University Activity

African universities and academic institutions are also setting up AI technology-based projects to tackle environmental issues, including Makerere University in Uganda and Carnegie Mellon University in Kigali, which was the first to offer a Master of Science Degree in Electrical and Computer Engineering with hands-on courses which include machine learning, robotics, and the internet of things (Carnegie Mellon University Africa). A few such projects are discussed below.

AirQo

The Makerere University Artificial Intelligence Research Group (AIR Lab) specializes in the application of artificial intelligence and data science to challenges common in the developing world. AIR Lab received support from the Pulse Lab to set up AirQo – an air quality data monitoring, analysis and modelling platform in East Africa meant to achieve clean air for all African cities through leveraging data (Nabatte 2019). AirQo is deploying a growing network of low-cost air quality monitors. Using machine learning and artificial intelligence to collect and analyze data, the project makes air quality predictions useful in raising awareness and informing policy decisions. Future research plans include the development and deployment of machine learning methodology to analyze air pollution data from Kampala, in order to determine the source of the pollution and to aid the design of mitigating interventions.

WIMEA-ICT

WIMEA-ICT is a combined research and capacity building project that seeks to improve weather information management in the entire East Africa region by development of ICT-based solutions (Norwegian Agency for Development Cooperation [NORAD] 2013). Funded by the Norwegian Agency for Development Cooperation (Norad) under the NORHED (Norwegian Programme for Capacity Development in Higher Education and Research for Development) scheme, the project is a cooperation between Makerere University in Uganda, Dar es Salaam Institute of Technology (DIT) in Tanzania, the University of Juba in South Sudan, and the Geophysical Institute of the University of Bergen.

The project recognizes the wide-ranging importance of weather data and the problems that result when weather predictions are inaccurate. Although project documentation does not specify the use of AI, among the five components of the project at least one is ideal for incorporation of AI: development of numerical weather prediction models specifically designed for the East African context.

The Potential of AI

It is not difficult to identify a long list of research projects that focus on various climate change issues in Africa. Many such projects include analyses of large datasets that would, seemingly, be ideal for analysis by AI algorithms. For example, Petja et al. (2004) describe an analysis of South African regional weather data dating from 1900 onward and satellite data dating from 1985 onward. The data are used to monitor regional climate and vegetation variations over time. In another example, Hagenlocher et al. (2014) describe the combination of numerous datasets to develop a cumulative climate change impact indicator. Applied to sub-Saharan Africa, the authors identified, evaluated, and mapped 19 hotspots that exhibited the most severe climate changes. In research out of Stanford University, Burke and Lobell (2017) demonstrated the importance of high-resolution satellite imagery data to estimate and understand yield variation among smallholder African farmers. This understanding generates various potential capabilities including the inexpensive measurement of the impact of specific interventions, the broader characterization of the source and magnitude of yield gaps, and the development of financial products aimed at African smallholders.

Although the immediately foregoing examples, and many other studies, do not specifically mention the use of AI, it is clear that large interrelated datasets are of vital importance to many different areas of research relevant to climate change. There is significant room for AI to be used by researchers to improve methodologies involving analysis of weather and other data (Rasp et al. 2018).

Challenges and Future Applications

Development of advanced technologies typically encounters challenges, and AI technologies and applications with a focus on climate change are no exception. Besides the typical challenges of developing AI products and services, Africa presents unique challenges both technological and political/ethical.

Scope of the Problem

Technologies powered by machine learning (ML) algorithms, including those discussed herein that aid climate analysis (Huntingford et al. 2019), have advanced dramatically, triggering breakthroughs in other research sectors. Although a considerable number of isolated Earth System features have been analyzed with ML techniques, more generic application to understand better the full climate system has not occurred, and the technology to do so may be quite far from the current state of development. At this stage of development, Artificial intelligence (AI) can be used to analyze smaller systems and provide enhanced warnings of approaching weather features, including extreme events. ML and AI can aid in understanding and improving existing data and simulations, as it has done in other systems (Huntingford et al. 2019). For instance, Airbus Defence and Space is using TensorFlow, the open-source set of AI tools from Google, to extract information from satellite images and offer valuable insights to customers. In a similar manner, AI can be used to detect and analyze isolated Earth System features and climate patterns, especially with the latest release of TensorFlow Quantum which enables a faster prototyping of ML models. Nevertheless, modeling the entire global climate remains challenging, and predictions from such models vary at all scales. Until the computational power and models have been refined to enable accurate global predictions, the need remains for smaller scale models. Local modeling requires context specific data and algorithms, so efforts toward development of Africa-specific climate change models must continue to be encouraged.

Ethical Issues of Predicting Climate Change Impacts

AI algorithms are well suited to analyze large datasets and detect patterns, so they are naturally well suited for looking at patterns of large-scale human movement and the data that might be associated with such movement (Beduschi 2020). For example, it is postulated that economic data such as GDP growth, along with trends in other data such as population growth and weather data (which might indicate food security issues), can be used to predict future large-scale human migration (Nyoni 2017). Accuracy of the predictions can be increased by incorporating real-time data points, such as announcements by government central banks, military actions, and weather observations.

Assuming there is a reasonable level of accuracy, predicting the location or the country most likely to suffer the next crisis of human migration has both remedial and prophylactic uses (TEDx Talks 2016). Humanitarian organizations can begin preparations for dealing with the crisis, and intergovernmental financial institutions can consider policy measures to ease debt burdens or encourage growth. Local and national governments of the yet-to-be affected regions can take measures to calm tensions and address the issues that cause migration. As useful as such predictions may be, this use case raises extreme ethical issues, for example by encouraging international efforts (including both active and passive efforts) to promote regime change where a specific government's policies appear to be leading to a future migratory crisis. Additionally, a prediction of a future migratory crisis may be a self-fulfilling prophecy, by increasing tension among the population and reducing investor confidence in the economy. The good intentions of those developing the technology, in this case, may increase the likelihood of the humanitarian disasters that they are seeking to ease.

Data Inadequacies

Apart from ethical issues, the development of AI for combating climate change in Africa may be severely hampered by a lack of data. The concept of a digital divide is many decades old, and documentation of the digital divide separating Africa from other regions is well established (Karar 2019). The modern-day extension of this concept is that of a data divide (Castro 2014), also referred to as a data desert or data poverty. First recognized with respect to certain populations in developed countries, the data divide is a problem in Africa and with respect to climate change for a variety of reasons. Historical weather data is less extensive in Africa compared with other parts of the world (Dinku 2018). Current data is also less extensive, as there are fewer weather satellites monitoring Africa than other regions and ground-based sensing is also less extensive (Dinku et al. 2011).

Even where there is historical climate data, those data may be inaccessible – for example, because African governments and their weather agencies are increasingly seeking to commercialize the data, or because the data are not digitized (Nordling 2019). In this context, the issue of a data divide is complex and is indicative of an uneven power dynamic. As with many other areas, Africa engages the rest of the world from a disadvantaged position, and the unbalanced power of the relationship may negatively affect the outcome. Whereas monetizing data is a common practice in developed countries, because African nations need significant help in building the infrastructure for collecting data, they are expected to willingly release the data. A data commons, in which climate data is readily available for all to use, is vitally important to help climate scientists and other interested parties understand the impacts of climate change in Africa. Nevertheless, the desire of data holders to seek ways of monetizing their data is understandable.

The concept of Africa as a data desert may, therefore, be unfairly characterizing the true situation. Rather than an absence of data, as the desert analogy implies, it is probably more accurate to say that data are present but not as readily available or easily searchable. As mentioned previously, African government websites may lack updated data (Ndongmo 2016), but this does not mean that they are not collecting the data. Open data portals are present in a few countries but have not become mainstream methods for governments to disseminate datasets. In some cases, governments generate revenue by selling datasets, and therefore have little incentive to making them available on an open platform. Efforts at increasing the volume of data collected about Africa should take into account these issues.

Regardless of the causes of the data divide, there is no doubt that insufficient data is important in Africa's ability to adapt to climate change. Climate change does not affect all geographical locations equally (United States Environmental Protection Agency (USEPA) 2017). As the global average temperature rises, and sea levels rise, average temperatures in some areas may decline. Overall rainfall may increase or decrease in any particular location, depending on a variety of factors. Extremes in precipitation and temperatures will also be in homogeneously affected. These variations would be less problematic if data were gathered with uniform consistency in all locations, but as mentioned previously, data collection in Africa is less consistent and less thorough. The result of this situation is likely to be less accurate predictions of the effects of climate change in Africa compared with other regions. Less accurate predictions may mean that local and international decision makers are unable to adequately prepare for the impacts of climate change.

Whether due to inefficient dissemination or to a fundamental lack of collection, or to some other reasons, the lack of available data has severe implications for the use of AI in adapting to climate change. Without sufficient data, AI algorithms are substantially less accurate and useful (West and Allen 2018). The trend in Africa, however, appears to be shifting toward a wider availability of data and a greater effort toward utilizing all available tools, including AI, in addressing climate change issues.

Conclusions

Some countries are serious in their look toward the future. For example, Ethiopia launched its first observatory satellite into space in 2019. The 70-kg remote sensing satellite is to be used for agricultural, climate, mining and environmental observations, allowing the Horn of Africa to collect data and improve its ability to plan for changing weather patterns for example. The satellite will operate from space around 700 km above the surface of earth. Developments in Ethiopia follow the introduction by the African Union of an African space policy, which calls for the development of a continental outer-space program and the adoption of a framework to use satellite communication for economic progress. Clearly, efforts such as this are forward thinking and will help the continent to address the lack of data that hampers the use of AI to address issues of climate change.

The problems of climate change are global, but Africa is likely to suffer to a greater degree compared with other regions. Scientists and policymakers in Africa need every available tool to help the continent adapt to the changes, but should always keep in mind the severity and scale of the problem. Notwithstanding the benefits that AI clearly brings, or promises to bring, in the efforts to adapt to climate change, it is clear that the fourth industrial revolution cannot fix what the second industrial revolution started.

References

Adeoye A (2019) Google has opened its first Africa Artificial Intelligence lab in Ghana. https://edition.cnn.com/2019/04/14/africa/google-ai-center-accra-intl/index.html. Accessed 16 Mar 2020

Adewumi AE (2020) AI for the people: how to stop the Nigerian energy crisis. https://omdena.com/blog/ai-energy-crisis. Accessed on 15 Mar 2020

African Soil Information Service, AfSIS. http://africasoils.net/. Accessed on 15 Mar 2020

Alliance for a Green Revolution in Africa (AGRA) (2014) Africa agriculture status report 2014: climate change and smallholder agriculture in sub-saharan Africa, AGRA Report. Nairobi. https://hdl.handle.net/10568/42343. Accessed on 16 Mar 2020

Alliance for a Green Revolution in Africa (AGRA) (2019) New partnership to boost food security in Africa by use of artificial intelligence. http://agra.org/new-partnership-to-boost-food-security-in-africa-by-use-of-artificial-intelligence. Accessed on 14 Mar 2020

Anderson K, Ryan B, Sonntag W, Kavvada A, Friedl L (2017) Earth observation in service of the 2030 agenda for sustainable development. Geo Spat Inf Sci 20(2):77–96. https://doi.org/10.1080/10095020.2017.1333230

Assefa S (2018) IBM Research, Hello Tractor pilot agriculture digital wallet based on AI and blockchain. https://www.ibm.com/blogs/research/2018/12/hello-tractor/. Accessed on 16 Mar 2020

Aydoğdu M, Boyraz N (2012) Stem rust (ug99), seen as a threat globally. Akden iz Univ Ziraat Fak Derg 25(1):23–28. https://www.researchgate.net/publication/305687716_Stem_rust_ug99_seen_as_a_threat_globally. Accessed on 16 Mar 2020

Beduschi A (2020) International migration management in the age of artificial intelligence. Migr Stud. https://doi.org/10.1093/migration/mnaa003

Burke M, Lobell DB (2017) Satellite-based assessment of yield variation and its determinants in smallholder African systems. Proc Natl Acad Sci U S A 114(9):2189–2194. https://doi.org/10.1073/pnas.1616919114

Campbell M, Tishkoff S (2008) African genetic diversity: implications for human demographic history, modern human origins, and complex disease mapping. Annu Rev Genomics Hum Genet 9:403–433. https://doi.org/10.1146/annurev.genom.9.081307.164258

Carnegie Mellon University Africa. https://www.africa.engineering.cmu.edu/. Accessed on 17 Mar 2020

Castro D (2014) The rise of data poverty in America. Report from Center for Data Innovation. Available at: http://www2.datainnovation.org/2014-data-poverty.pdf. Accessed on 13 Mar 2020

Chiwewe T, Ditsela J (2016) Machine learning based estimation of ozone using spatio-temporal data from air quality monitoring stations. Paper presented at the 14th International Conference on Industrial Informatics, Poitiers, France, 18–21 July 2016. https://doi.org/10.1109/INDIN.2016.7819134

Cox H, Kelly K, Yetter L (2014) Using remote sensing and geospatial technology for climate change education. J Geosci Educ 62(4):609–620. https://doi.org/10.5408/13-040.1

Deep Learning Indaba. http://www.deeplearningindaba.com. Accessed on 16 Mar 2020

Department of Environmental Affairs South Africa (2016) Welcome to environmental GIS. https://egis.environment.gov.za/. Accessed on 16 Mar 2020

Diallo AB, Nguifo EM, Dhifli W, Azizi E, Prabhakaran S, Tansey W (2019) Selected papers from the workshop on computational biology: joint with the international joint conference on artificial intelligence and the international conference on machine learning 2018. J Comput Biol 26:6. https://doi.org/10.1089/cmb.2019.29020.abd

Diamandis P (2019) AI and robotics are transforming disaster relief. https://singularityhub.com/2019/04/12/ai-and-robotics-are-transforming-disaster-relief/. Accessed on 17 Mar 2020

Dinku T (2018) Overcoming challenges in the availability and use of climate data in Africa. ICT Update. https://ictupdate.cta.int/en/article/overcoming-challenges-in-the-availability-and-use-of-climate-data-in-africa-sid06fd8a811-e179-4fa5-9c8f-806bd2f27c3e. Accessed on 13 Mar 2020

Dinku T, Asefa K, Hilermariam K, Grimes D, Connor S (2011) Improving availability, access and use of climate information. World Meteor Org Bull 60(2) Available at: https://public.wmo.int/en/bulletin/improving-availability-access-and-use-climate-information. Accessed on 16 Mar 2020

European Commission High-Level Expert Group on Artificial Intelligence (2018) The ethics guidelines for trustworthy artificial intelligence. https://ec.europa.eu/digital-single-market/en/high-level-expert-group-artificial-intelligence. Accessed on 16 Mar 2020

GhanaWeb (2019) GSS partners Vodafone & Flowminder to produce reliable data for sustainable development. https://www.ghanaweb.com/GhanaHomePage/NewsArchive/GSS-partners-

Vodafone-amp-Flowminder-to-produce-reliable-data-for-sustainable-development-814189. Accessed on 17 Mar 2020

Hagenlocher M, Lang S, Hölbling D, Tiede D, Kienberger S (2014) Modeling hotspots of climate change in the Sahel using object-based regionalization of multidimensional gridded datasets. IEEE J Sel Top Earth Obs Remote Sens 7(1):229–234. https://doi.org/10.1109/JSTARS.2013.2259579

Huntingford C, Jeffers E, Bonsall M, Christensen H, Lees T, Yang H (2019) Machine learning and artificial intelligence to aid climate change research and preparedness. Environ Res Lett 14 (12):124007. https://doi.org/10.1088/1748-9326/ab4e55

International Telecommunication Union (2019) United Nations activities on artificial intelligence. Compendium presented and discussed at the AI for Good UN Partners Meeting, UNWomen Headquarters in New York, 23 September 2019. https://www.itu.int/dms_pub/itu-s/opb/gen/S-GEN-UNACT-2019-1-PDF-E.pdf. Accessed on 16 Mar 2020

Johnson K (2018) Major AI conference is moving to Africa in 2020 due to visa issues. https://venturebeat.com/2018/11/19/major-ai-conference-is-moving-to-africa-in-2020-due-to-visa-issues/. Accessed on 16 Mar 2020

Johnson K, Pasquale F, Chapman J (2019) Artificial intelligence, machine learning, and bias in finance: toward responsible innovation. Fordham Law Rev 88(2):499. https://ir.lawnet.fordham.edu/flr/vol88/iss2/5. Accessed on 12 Mar 2020

Karar H (2019) Algorithmic capitalism and the digital divide. J Dev Soc 35(4):514–537

Knight W (2019) African AI experts get excluded from a conference – again. https://www.wired.com/story/african-ai-experts-get-excluded-from-a-conference-again/. Accessed on 16 Mar 2020

Lecoq M (2003) Desert locust threat to agricultural development and food security and FAO/International role in its control. Arab J Pl Prot 21:188–193. https://www.researchgate.net/publication/242118174_Desert_Locust_Threat_to_Agricultural_Development_and_Food_Security_and_FAOInternational_Role_in_its_Control. Accessed on 17 Mar 2020

Lukhanyu M (2019) Agri SA & Aerobotics partner to offer free service for all South African farmers. https://techmoran.com/2019/05/15/agri-sa-aerobotics-partner-to-offer-free-service-for-all-south-african-farmers/. Accessed on 16 Mar 2020

Meier P (2012) Crisis mapping in action: how open source software and global volunteer networks are changing the world one map at a time. J Map Geogr Libr 8(2):89–100. https://doi.org/10.1080/15420353.2012.663739. Accessed on 15 Mar 2020

Microsoft Access Partnership University of Pretoria (2018) Artificial intelligence for Africa: An opportunity for growth, development and democratisation. White Paper. https://www.up.ac.za/media/shared/7/ZP_Files/ai-for-africa.zp165664.pdf. Accessed on 14 Mar 2020

Nabatte P (2019) Mak's AirQo project is a US 1.3million Google AI impact grantee. https://news.mak.ac.ug/2019/05/maks-airqo-project-us13m-google-ai-impact-grantee. Accessed on 14 Mar 2020. See further AirQo https://www.airqo.net/. Accessed on 17 Mar 2020

Nanalyze (2019) Top-10 artificial intelligence startups in Africa. https://www.nanalyze.com/2019/04/artificial-intelligence-africa/. Accessed on 16 Mar 2020

Ndongmo K (2016) African government websites – new and improved, or not? Report on medium. Edition of January 8 2016. Available at: https://medium.com/@kathleenndongmo/african-government-websites-new-and-improved-or-not-746a5a5a62d. Accessed on 16 Mar 2020

Nordling L (2019) Scientists struggle to access Africa's historical climate data. Nature 574:605–606

Norwegian Agency for Development Cooperation (NORAD) (2013) Improving weather information management in East Africa. https://norad.no/en/front/about-norad/. Accessed 10 Mar 2020

Nyoni B (2017) How Artificial Intelligence can be Used to Predict Africa's Next Migration Crisis. https://www.unhcr.org/innovation/how-artificial-intelligence-can-be-used-to-predict-africas-next-migration-crisis/. Accessed on 13 Mar 2020

Nzouankeu A (2019) App helps African farmers detect crop disease. https://www.voanews.com/africa/app-helps-african-farmers-detect-crop-disease. Accessed on 17 Mar 2020

Omdena (2019) Using AI to predict climate change and forced displacement. https://omdena.com/challenges/ai-climate-change. Accessed on 10 Mar 2020

One Concern (2020) What we do: We are advancing science based technology to make disasters less disastrous. https://www.oneconcern.com/product. Accessed on 16 Mar 2020

Oxford Insights International Development Research Centre (2019) Government artificial Intelligence readiness index 2019. https://www.oxfordinsights.com/ai-readiness2019. Accessed on 12 Mar 2020

Patrick S (2020) How should the world respond to the coming wave of climate migrants? World Politics Review. https://www.worldpoliticsreview.com/articles/28603/how-should-the-world-respond-to-the-coming-wave-of-climate-migrants. Accessed on 15 Mar 2020

Peacock C, Mungai F (2019) Impact investment favours expats over African entrepreneurs: Here's how to fix that. World Economic Forum. https://www.weforum.org/agenda/2019/07/impact-investors-favour-expats-over-african-entrepreneurs-here-s-how-to-fix-that/. Accessed on 12 Mar 2020

Penn State. (2019). New AI app predicts climate change stress for farmers in Africa: researchers will unveil their app for climate-smart agriculture to coincide with the UN Climate Action Summit. http://www.sciencedaily.com/releases/2019/09/190923082252.htm. Accessed 12 Mar 2020

Perrault R, Shoham Y et al (2019) The AI index 2019 annual report. AI Index Steering Committee Human-Centered AI Institute, Stanford University, Stanford. Accessed on 13 March 2020. Available at: https://hai.stanford.edu/sites/xg/files/sbiybj10986/f/ai_index_2019_report.pdf

Peters A (2018) This startup lets african farmers hire and on-demand tractor to boost their harvests. https://www.fastcompany.com/90227534/hello-tractor-and-john-deere-bring-10000-tractors-to-africa. Accessed on 20 Sept 2019

Petja BM, Malherbe J, van Zyl D (2004) Using satellite imagery and rainfall data to track climate variability in South Africa. Paper presented at the 2004 IGARSS 2004 IEEE International Geoscience and Remote Sensing Symposium, Anchorage, Alaska, 20–24 September 2004. https://doi.org/10.1109/IGARSS.2004.1369098

Pulse Lab Kampala (2018) Pulse Lab Kampala progress report 2016–2017. https://beta.unglobalpulse.org/document/pulse-lab-kampala-progress-report-2016-2017-2/. Accessed on 16 Mar 2020

Rasp S, Pritchard MS, Gentine P (2018) Deep learning to represent subgrid processes in climate models. Proc Natl Acad Sci U S A 115(39):9684–9689. https://doi.org/10.1073/pnas.1810286115

Scoones I, Thompson J (2011) The politics of seed in Africa's green revolution: alternative narratives and competing pathways. IDS Bull 42(4):1–23

Starbird K (2012) Crowdwork, crisis and convergence: how the connected crowd organizes information during mass disruption events. Dissertation, University of Colorado. http://faculty.washington.edu/kstarbi/starbird_dissertation_final.pdf. Accessed on 14 Mar 2020

TEDx Talks (2016) Predicting Africa's Next Refugee Crisis Using AI|Babusi Nyoni| TEDxCapeTown In: Youtube. https://www.youtube.com/watch?v=lsi8ZgTvbT4. Accessed on 14 Mar 2020

The Economist (2020) China's success on AI has relied on good data. https://www.economist.com/technology-quarterly/2020/01/02/chinas-success-at-ai-has-relied-on-good-data. Accessed on 10 Mar 2020

The Guardian (2017) 'Tsunami of data' could consume one fifth of global electricity by 2025. https://www.theguardian.com/environment/2017/dec/11/tsunami-of-data-could-consume-fifth-global-electricity-by-2025. Accessed on 10 Mar 2020

The World Bank (2014) Land and food security. https://www.worldbank.org/en/topic/agriculture/brief/land-and-food-security1. Accessed on 17 Mar 2020

TOP500.org (2019) TOP500 lists. https://www.top500.org/lists/top500/. Accessed on 6 Mar 2020

Tortoise Intelligence (2019) The global AI index. https://www.tortoisemedia.com/intelligence/ai/. Accessed on 12 Mar 2020

United Nations World Data Forum (2018) (TA2.16) Earth observation applications for the sustainable development goals: opportunities for scaling successful methods. https://unstats.un.org/

unsd/undataforum/sessions/ta2-16-earth-observation-applications-for-the-sustainable-develop
ment-goals-opportunities-for-scaling-successful-methods/. Accessed on 16 Mar 2020

United States Environmental Protection Agency (2017) International Climate Change Impacts.
Report from Climate Change Impacts edition of January 19 2017. Available at: https://19januar
y2017snapshot.epa.gov/climate-impacts/international-climate-impacts_.html. Accessed on 16
Mar 2020

Vodafone (2018) Vodafone Foundation to use aggregated anonymised mobile data to help prevent
spread of epidemics in transformative 'Big data for good' programme. https://www.vodafone.
com/news-and-media/vodafone-group-releases/news/vodafone-foundation-use-aggregate-data.
Accessed on 15 Mar 2020

Weldemariam K, Pickover C, Kozloski J, Gordon M (2020) Artificial Intelligence for providing
enhanced microblog message insertion, US Patent Application 20200028885, 23 Jan 2020

West D, Allen J (2018) How Artificial Intelligence is transforming the world. The Brookings
Institution report. Available at: https://www.brookings.edu/research/how-artificial-intelligence-
is-transforming-the-world/. Accessed on 04 Nov 2020

3

Differential Impact of Land Use Types on Soil Productivity Components in Two Agro-Ecological Zones of Southern Ghana

Folasade Mary Owoade, Samuel Godfried Kwasi Adiku, Christopher John Atkinson and Dilys Sefakor MacCarthy

Contents

Introduction ....
Materials and Methods
   Sites
   Sampling and Analysis
   Statistical Analysis
Results and Discussions
   Effect of Site Locations on Soil Properties
Conclusion
References

F. M. Owoade (✉)
Department of Crop Production and Soil Science, Ladoke Akintola University of Technology, Ogbomoso, Nigeria
e-mail: fmowoade@lautech.edu.ng

S. G. K. Adiku
Department of Soil Science, University of Ghana, Legon, Ghana
e-mail: s_adiku@ug.edu.gh

C. J. Atkinson
Natural Resources Institute, University of Greenwich, London, UK

Department of Agriculture, Health and Environment, Natural Resources Institute, University of Greenwich, Chatham, UK
e-mail: C.J.Atkinson@greenwich.ac.uk

D. S. MacCarthy
Soil and Irrigation Research Centre Kpong, University of Ghana, Accra, Ghana
e-mail: dsmaccarthy@gmail.com

Abstract

The maintenance of soil productivity is important for sustained crop yield in low-input systems in the tropics. This study investigated the impact of four different land use types, namely, maize and cassava cropping, woodlot/plantations, and natural forests on soil productivity components, especially soil carbon accretion, at six sites within two agro-ecological zones of southern Ghana. Soil properties were significantly different between sites and ecological zones. The coastal savanna zones, which is a low rainfall zone had relatively lower soil carbon storage than the high rainfall forest-savanna transition zone. Soil productivity conditions in the later zone were much more favorable for cropping than the former. Land use types significantly affected the soil carbon (*SOC*) storage within the two ecological zones. In the low rainfall zone, soil carbon accretion by maize cropping, cassava cropping, and plantations were 48%, 54%, and 60%, respectively, of the forest carbon stock (47,617 kg/ha). In the transition zone, the soil carbon accretion was over 90% of the forest value (48,216 kg/ha) for all land use types. In effect use of land use types in maintaining soil productivity must consider the conditions in a given ecological zone.

Keywords

Agro-ecology · Land use · Soil carbon stock · Soil productivity · Soil properties

Introduction

Soil organic carbon (*SOC*) is a major component of productivity in low-input cropping systems of the tropics. Soil carbon influences the physical, chemical, and biological properties of the soil. Many studies have indicated that the reduction in the *SOC* can result in significant decrease in the available water capacity (Hudson 1994), structural deterioration, and an increased bulk density (Shu et al. 2015). Also, the contribution of the *SOC* to soil fertility maintenance is also well established. Crop yield reduction is often associated with *SOC* losses, largely because the *SOC* is a major reservoir of nutrients, especially in the tropics where external inputs continue to remain low (Sanchez et al. 2009). Estimates by Lal (2006) indicated that maize yield could decline by 30–300 kg ha^{-1} for every ton ha^{-1} of *SOC* in the root zone. Regarding soil biology, the *SOC* is a major source of nutrition and energy for microbial life. Some authors describe the *SOC* as the "... life blood" of tropical soils (Acquaye 1989). Though the *SOC* plays a dominant role in tropical agriculture, other soil properties may also enhance the overall productivity. Nutrients elements such as nitrogen, which is largely derived from organic matter mineralization, phosphorus from rock minerals, and the overall cation retention capacity are important factors that also determine soil and crop productivity.

Soil carbon and hence productivity is not permanent but may change rapidly depending on land use type and management (Zerihun 2017; Waddington et al.

2010; Reynolds et al. 2015). Much of the literature (Burras et al. 2001; Sa et al. 2001; Batlle-Aguilar et al. 2011) indicates that the conversion of forest to agriculture and other forms of land use such as plantations and woodlots is the major cause of soil productivity decline in the tropics. Brams (1971) showed a 50% reduction in the *SOC*, only 5 years after forest clearing in Sierra Leone.

Residue management methods employed in agriculture also lead to changes in *SOC*. Adiku et al. (2009) showed in Ghana that where crop residues were removed (e. g., by burning, or cutting to feed animals with no return of manure), the *SOC* declined rapidly from the long-term fallow land value of 18 g kg^{-1} to 7 g kg^{-1} with 4 years of maize cropping. However, where the residues were maintained as mulch, the rate of *SOC* decline was much slower, from 18 g kg^{-1} to 15 g kg^{-1} over years. A greater buildup of SOC in forests than other land use types would be expected because of long-term continuous litter addition (Brinson et al. 1980), for example, by avoidance of cultivation losses and reduced decomposition due to lower temperatures under a tree canopy. For croplands, the constant disturbance of the soil enhances *SOC* decomposition (Lal 1997; Hulugalle et al. 1984; Kang 1993; Dalal et al. 1991), and the constant harvest or removal of plant organic material (Feller 1993) would increase the *SOC* loss due to the partial exposure of the soil to high temperatures during off-seasons.

Despite these findings, the manner in which land use types affect soil carbon storage in different ecological zones is not well understood. The question of interest here is whether a given land use type will equally impact soil properties in different rainfall and vegetation zones. In other words, can we generalize that cropping will adversely affect soil productivity irrespective of the carbon input capacity of different ecological zones? This aspect of research is still lacking in Ghana, even though it has relevance for the design and management of soil productivity. The focus of this chapter is to examine how four land use types (forest, woodlot/plantation, cassava cropping, and maize cropping) affect soil carbon content and other properties at six farming sites of Ghana (across two ecological zones).

Materials and Methods

Sites

Six (6) farming sites from two agro-ecological zones, all in southern Ghana, were selected for this study (Fig. 1). Three of the farming sites (Accra Metropolis, Ga East District, Ga West District) fall in the coastal savanna zone of Ghana and receives 650–1000 mm rainfall. Though the vegetation is largely grassland, some derived savanna locations still host original pockets of forestland. The dominant soil type of the Greater Accra Region based on FAO/UNESCO classification is Ferric Acrisol and Umbic Leptosol (Soil Research Institute 1999). The remaining three sites (Yilo Krobo District, Shai Osudoku District, and Upper Manya District) fall within the forest-savanna transition zone (hereinafter transition zone) receiving 1500–2000 mm rainfall. Vegetation is largely forest at some portions and mixed with grassland in other portions. The soils of the transition comprise Cambic Arenosol and Calcic

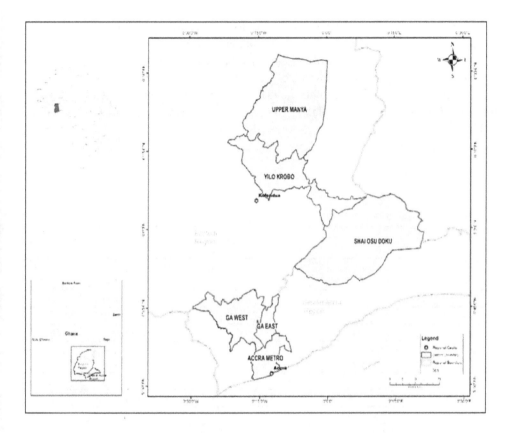

Fig. 1 Map shows position of Ghana in West Africa (top left), the location of the study sites in southern Ghana (bottom left), and the details of the districts (Ga West, Ga East, Accra Metropolis, Yilo Krobo, Shai Osudoku, and Upper Manya) used in Ghana

vertisol. Rainfall in both ecological zones is bimodally distributed, with a major wet season from March to July and a minor wet season from September to November. Agriculture is the main source of livelihood at all the sites.

The first site (Accra Metropolis) in the coastal savanna zone was located at the University of Ghana Farm and hosts a forest of more than 150 years old. Nearby the forest is cleared area which has been cropped to maize (*Zea mays*) and cassava (*Manihot esculenta*) for more than 20 years by University Farm workers. The maize fields receive periodically modest fertilizer application of not more than 30 kg N ha^{-1}. Also, located at the University Farm is an 18-year-old woodlot of *Leucaena species* established on a previously cropped land. Experimental plots at the University Farm were not included in this study. The soils under the forests and woodlots have little mechanical disturbance for many years, but the croplands are plowed and cultivated annually. The second site in this zone (Abokobi) also carried a 70-year-old forest with adjacent lands cropped to maize and cassava for not more than 10 years. A plantation of plantain (*Musa* spp.) was established near the forest. The third site (Pokuase) carried a forest of more than 70 years, an oil palm (*Elaeis guineensis*) plantation as well as cassava and maize farms.

   The fourth site (Yilo Krobo), which is located in the transition zone carried a forest of more than 50 years, an oil palm plantation and relatively young crop farms of maize and cassava. The fifth site (Shai Osudoku) holds a protected forest with mature trees of more than 100 years old. For this reason, soil sampling was restricted to the fringes of the forest. Other land use systems at this site include an 18-year mango (*Mangifera indica*) plantation, maize, and cassava farms. The sixth site (Upper Manya) has a forest of more than 50 years old, along with oil palm plantations, and recent arable farms.

Table 1 Land use types at the two agroecological zones

Site	Ecology	Rainfall (mm)	Dominant soil	Land use	Age (years)
Accra-Metropolis	Coastal savannah	700	Ferric Acrisol	Maize	10
				Cassava	10
				Woodlot	20
				Forest	>150
Ga-East	Coastal savannah	800	Ferric Acrisol	Maize	5
				Cassava	8
				Plantation	6
				Forest	>70
Ga-West	Coastal savannah	800	Umbic Leptosol	Maize	4
				Cassava	10
				Plantation	10
				Forest	>70
Yilo Krobo	Forest savannah	900–;1500	Calcic Vertisol	Maize	10
				Cassava	8
				Plantation	23
				Forest	>60
Shai-Osudoku	Forest savannah	900–1500		Maize	>50
				Cassava	>50
				Plantation	18
				Forest	>100
Upper-Manya	Forest savannah	900–1500	Cambic Arenosol	Maize	5
				Cassava	7
				Plantation	10
				Forest	>50

Farmers' best estimates

Sampling and Analysis

For this study, four land use types, namely, maize cropping, cassava cropping, plantations (teak, mango, plantain, oil palm, woodlot), and natural forests were selected (Table 1). Fifty-four (54) farms across the six sites were selected and visited from April to May 2017, and soils were sampled from each land use type from the depth of 0–20. The cropping history of the 54 farms were documented during a prior survey by Owoade et al. (2017). At each site, at least three maize and three cassava farms were sampled in triplicates and bulked to obtain a composite sample for each farm. In addition, soils were sampled from the plantations and the natural forests.

The disturbed top soil (0–20 cm) and separately sampled undisturbed soil cores were brought to the laboratory for analysis. The bulk density was determined on the soil cores. Disturbed soils were air-dried, crushed, and sieved through 2-mm sieve for the determination of texture, total soil carbon, pH, total nitrogen, available phosphorus, and exchangeable cations (K, Mg, Ca, Na). Exchangeable bases were determined by extraction with 250 ml of buffered 1.0 M ammonium acetate followed by flame photometric determination and the effective cation exchange capacity (ECEC) was determined as the sum of exchangeable cations. Soil texture determination followed the procedure of Bouyoucous (1951) as modified by Day (1965) using sodium hexametaphosphate as the dispersant. Soil pH was determined in 1:1 soil to water ratio using a MV88 Praitronic pH meter and electrode. Available phosphorus (AvP) was determined colorimetrically after extraction with Bray 1 solution (Bray and Kurtz 1945) and the concentration measured using a UV-Spectrophotometer. Total soil carbon and nitrogen were determined using TruMac Carbon, Nitrogen, and Sulfur analyzer (Model N1914).

The soil carbon content was converted to stocks (C_{st}) using:

$$C_{st} = A \times \rho_b \times z \times SOC \tag{1}$$

where A is the land area (1 ha $= 10^4$ m^2), ρ_b is the soil bulk density (kg/m^3), and z is the soil depth (0.20 m).

Statistical Analysis

Analysis of variance was conducted using MINITAB software to determine the influence of land use types and site location on soil properties.

Table 2 Variation of soil properties with sites

Site	Ecology	Bulk density (g/cm³)	SOC (g/kg)	pH	Avail P (mg/kg)	N (mg/kg)	ECEC (cmol/kg)
Accra Metropolis	Coastal savanna	1.27 ± 0.16	12.4 ± 6.0	4.4 ± 0.4	18.7 ± 12	4.3 ± 0.4	3.11 ± 0.86
Ga East		1.38 ± 0.10	11.49 ± 5.0	5.05 ± 0.5	10.9 ± 6.0	4.1 ± 0.3	3.06 ± 0.91
Ga West		1.29 ± 0.14	15.6 ± 8.6	5.6 ± 0.54	10.8 ± 6.1	1.90 ± 0.9	4.12 ± 1.4
Yilo Krobo	Transition	1.19 ± 0.15	23.50 ± 6.5	6.15 ± 0.54	15.70 ± 4.3	1.20 ± 0.6	6.79 ± 2.16
Shai Osudoku		1.32 ± 0.07	18.20 ± 5.8	6.15 ± 0.39	12.90 ± 7.1	3.6 ± 2.2	6.64 ± 1.52
Upper Manya		1.29 ± 0.08	13.10 ± 2.7	6.03 ± 0.0.48	7.90 ± 4.2	1.80 ± 0.23	4.29 ± 1.5
p-value		0.007	0.000	0.000	0.008	0.000	0.000

Results and Discussions

Effect of Site Locations on Soil Properties

There were significant differences among some soil properties at the various sites (Table 2). For example, soil texture, expressed as clay ratio (sand + silt)/clay), differed significantly with site, with high values (>11) at sites 1, 2, 5, and 6, and low values (4–5) at locations 3 and 4 (not shown). The bulk density differed significantly with site (p = 0.007) with higher values observed for the coastal savanna zone (1.3–1.4 g/cm^3) than in the transition zone (1.2–1.3 g/cm^3). The SOC also differed significantly with site (p = 0.000), with higher values in the transition zone (1.3–2.4 g/kg) than the coastal savanna zone (1.2–1.6 g/kg). Apparently, the higher SOC of the transition zone could be attributed to a greater carbon input by the high rainfall and more forest vegetation than the savanna ecological zone. The higher SOC of the transition zone may explain, the lower bulk density values, as these two properties are inversely related.

The pH differed significantly (p = 0.000) between ecological zones with the transition zone having higher values (6.0–6.2) than the coastal savanna (4.4–5.6). Soil pH differences may have consequences to crop performance because most nutrient elements are usually available in the pH range of 5.5–6.5 (Motsara and Roy 2008). The available P was significantly different (p = 0.008) among the sites. Though site 1 (Accra Metropolis) in the coastal savanna zone had the highest value (18.7 mg/kg), and the variability was also very high (SD = 12 mg/kg). Except for the site 6 (Upper Manya) which had the lowest available P (7.9 mg/kg), the average P values were quite similar for the two ecological zones; 11–18 mg/kg for the coastal savanna and 8–16 mg/kg for the transition zone. In general, the available P levels were somewhat adequate for plant growth, given a threshold of 11 mg/kg (Adeoye and Agboola 1985). Total nitrogen values varied significantly with site (p = 0.000) with higher values for the coastal savanna (1.1–4.3 mg/kg) than the transition zone (1.2–3.6 mg/kg). Though the ECEC can be considered as generally low but differed significantly (p = 0.000) between ecological zones, with the transition zone having 4.3–6.8 cmol/kg and the coastal savanna having 3.1–4.1 cmol/kg. Based on the soil property values, it may be concluded that the transition zone provided a much better soil condition for cropping than the coastal savanna zone.

Not all the soil properties were significantly affected by land use type. Across all sites, land use type had significant (p = 0.003) effect on the SOC, with a clear-cut difference between the SOC of the forest (21 g/kg) and the rest of the land use types where the SOC ranged from 13.0 (maize) to 14.0 (woodlot/plantation) g/kg (Table 3). Land use type also had significant effect (p = 0.000) on the bulk density, with the forests having the lowest value (1.2 g/cm^3) and maize farms having the highest (1.34 g/cm^3). There were no significant effects of land use types on available P, N, ECEC, or pH.

The interactive effects between sites and land use type were only significant for the SOC and bulk density. Incidentally, these two properties are the major determinants of the total carbon stock (Eq. 1). Land use type significantly (p = 0.013) affected the total carbon stocks at the various sites in the order: forest > plantation > cassava > maize. Across the sites, the forest soils had the highest average storage of 48,216 ± 12,811 kg/

Table 3 Variation of soil properties with land use types

Land use	Bulk density (g/cm^3)	SOC (g/kg)	pH	Avail P (mg/kg)	N (mg/kg)	ECEC (cmol/kg)
Maize	1.34 ± 0.15	13.0 ± 7.40	5.6 ± 0.79	14.8 ± 7.7	2.50 ± 1.3	4.60 ± 0.1.6
Cassava	1.29 ± 0.08	15.00 ± 6.4	5.70 ± 0.82	11.61 ± 7.0	2.90 ± 1.4	4.60 ± 2.4
Plantation	1.32 ± 0.14	14.0 ± 5.90	5.60 ± 0.82	11.80 ± 6.3	3.00 ± 1.3	4.40 ± 1.7
Forest	1.20 ± 0.10	20.9 ± 6.9	5.30 ± 0.71	13.10 ± 10	2.70 ± 2.3	5.10 ± 2.5
p-value	0.005	0.003	NS	NS	NS	NS

Table 4 Land use effects on soil carbon stocks

Region	Land use	Carbon stock (kg/ha)	% Forest carbon
Coastal savanna			
	Maize	24,316 ± 12,155	48
	Cassava	27,671 ± 10,839	54
	Woodlot	30,325 ± 4450	60
	Forest	50,491 ± 12,755	–
Forest-savanna transition			
	Maize	41,897 ± 12,263	88
	Cassava	43,001 ± 10,007	91
	Plantation	45,019 ± 14,786	95
	Forest	47,382 ± 20,252	–

ha "..(not shown) 4" followed by plantations (36,774 ± 14,482 kg/ha), cassava (35,007 ± 14,014 kg/ha), and maize (33,905 ± 12,811 kg/ha). Other studies (Djagbletey et al. 2018) have also reported higher soil carbon stocks for denser forests in the Guinean savanna zone of Ghana. In other works, carbon stocks as high as 59,450 kg/ha determined for forest soils in the semi-deciduous forest zone of Ghana by Dawoe (2009).

With regard to ecological zones, the results showed that the differences in land use impact on soil carbon in the transition zone was smaller than in the coastal savanna zone (Table 4). In the coastal savanna zone, maize accrued 48% of the forest carbon stock, while the cassava and plantation accrued 54% and 60%, respectively, suggesting that the plantations were most effective in soil carbon restoration. With regard to the transition zone, however, the soil carbon restoration effectiveness was generally high 90–95% for all the land use types. Though the plantations impact was again the highest, values above 90% accretion generally suggests that all the land use types were equally effective.

Options for *SOC* accretion must consider the differential effects of land use types as well as ecological zones. Our observations indicated that cropping of the land depleted the *SOC* the most. The reduction of *SOC* stocks on cropped lands can be attributed to factors such as the harvest removal of plant organic matter (Feller 1993), constant disturbance of the soil that enhances decomposition (Lal 1999; Hulugalle et al. 1984; Kang 1993; Dalal et al. 1991), and the partial exposure of soil to high temperatures during off-seasons when vegetation cover is reduced. These researchers (Yilfru and Taye 2011; Caravaca et al. 2002; Malo et al. 2005) also recorded greater SOC in forest compared to cultivated land. Though our sampling did not include intercropping systems, the observation that cassava cropping accrued high carbon stock than maize suggests that a maize-cassava intercropping could, perhaps, maintain a higher carbon addition, because of the longer life cycle of cassava and higher SOC accretion than maize.

The plantation land use type showed a higher SOC accretion but would require relatively long periods of time to achieve, thus preventing cropping of the lands for some time. Presumably, a combination of plantation and cropping is desired to ensure both crop productivity as well as *SOC* maintenance. This can be accomplished by agroforestry, which has been promoted in many parts of the tropics to enhance crop yields (Kang 1993), but the adoption rates have continued to be very low. Apparently, the competition between live trees and crops for resources in agroforestry systems can reduce crop productivity (Ong et al. 1991), thereby handicapping the adoption of the system by farmers. Tree-crop rotations, as practiced under the traditional shifting cultivation, could also be an effective alternative to agroforestry. The tree phase of the rotation would permit the rebuild of the *SOC* which is depleted during the cropping phase. This traditional shifting cultivation needs to be further researched for its role in soil carbon management and crop production.

Conclusion

Soil productivity in Ghana is influenced by both land use type and agro-ecological zone. Findings from this study indicated that soil productivity conditions for agriculture were less favorable for agriculture in the coaster savanna than the forest-savanna transition zone. Furthermore, land use types had significant impact on the carbon storage, with maize-based cropping systems having the lowest carbon stocks. Woodlot/plantation types of land use restored the SOC and productivity more effectively than croplands. In effect, the effectiveness of land use systems for soil productivity maintenance differs with agro-ecological zones. This must be factored into the design of land management measures.

Acknowledgment This research is supported by funding from the UK's Department for International Development (DfID) under the Climate Impacts Research Capacity and Leadership Enhancement (CIRCLE) program implemented by the African Academy of Sciences and the Association of Commonwealth Universities.

References

Acquaye DK (1989) Towards the development of sustainable agriculture in Africa. Reflections of a soil scientist. In: Proceedings of the 11th and 12th general meeting, Soil Science Society of Ghana. University of Ghana, Accra, p 91

Adeoye GO, Agboola AA (1985) Critical levels for soil pH, available P, K, Zn and Mn and maize ear-leaf content of P, Cu and Mn in sedimentary soils of South-Western Nigeria. Nutr Cycl Agroecosyst 6:65–71

Adiku SGK, Jones JW, Kumaga FK, Tonyiga A (2009) Effects of crop rotation and fallow residue management on maize growth, yield and soil carbon in a savanna- forest transition zone of Ghana. J Agric Sci 147(3):313–322

Batlle-Aguilar B, Porporato A, Barry D (2011) Modelling soil carbon and nitrogen cycles during land use change. A review. Agron Sustain Dev 31(2):251–274

Bouyoucous GH (1951) A calibration of the hydrometer for making mechanical analysis of soils. Agron J 43:434–438

Brams EA (1971) Continuous cultivation of West African soils: organic matter diminution and effects of applied lime and phosphorus. Plant Soil 35:401–414. https://doi.org/10.1007/BF01372671

Bray RH, Kurtz LT (1945) Determination of total, organic, and available forms of phosphorus in soil. Soil Sci 59:39–45

Brinson M, Bradshaw HD, Holmes RN, Elkins JB Jr (1980) Litterfall, stemflow and throughfall nutrient fluxes in an alluvial swamp forest. Ecology 61:827–835

Burras L, Kimble JM, Lal R, Mausbach MJ, Uehara G, Cheng HH, Kissel DE, Luxmoore RJ, Rice CW, Wilding LP (2001) Carbon sequestration: position of Soil Science Society of America. http://www.soils.org/carbseq.html

Caravaca F, Masciandaro G, Ceccanti B (2002) Land use in relation to soil chemical and biochemical properties in a semi-arid Mediterranean environment. Soil Tillage Res 69(1):23–30. https://doi.org/10.1016/s0167-1987(02)00080-6

Dalal RC, Strong WM, Weston EJ, Gaffney J (1991) Soil fertility decline and restoration of cropping lands in sub-tropical Queensland. Trop Grasslands 25:173–180

Dawoe E (2009) Conversion of natural forest to cocoa agroforest in Lowland Humid Ghana: impact on plant biomass production, organic carbon and nutrient dynamics. PhD thesis, Department of Agroforestry, Kwame Nkrumah University of Science and Technology, Kumasi, 260 pp

Day PR (1965) Particle fractionation and particle size analysis. In: Black CA et al (eds) Methods of soil analysis, Part I. Agronomy, vol 9. American Society of Agronomy, Madison, pp 545–567

de Sa JCM, Cerri CC, Dick WA, Lal R, Venske-Filho SP, Piccolo MC, Feigl BE (2001) Organic matter dynamics and carbon sequestration rates for a tillage chronosequence in a Brazilian Oxisol. Soil Sci Soc Am J 65:1486–1499

Djagbletey ED, Logah V, Ewusi-Mensah N, Tuffuour HO (2018) Carbon stocks in the Guinea savanna of Ghana: estimates from three protected areas. Biotropica 1–9. https://doi.org/10.1111/btp.12529...org/10.5061/dryad.d1d30

Feller OE (1993) Organic input, soil organic matter and functional soil organic compartments in low-activity clay soils in tropical zones. In: Mulongoy K, Merckx R (eds) Soil organic matter dynamics and sustainability of tropical agriculture. Wiley, New York, pp 77–85

Hudson BD (1994) Soil organic matter and available water capacity. J Soil Water Conserv 49:189–194

Hulugalle NR, Lal R, Ter Kuille CHH (1984) Soil physical changes and crop root growth following different methods of land clearing in Western Nigeria. Soil Sci 138:172–179

Kang BT (1993) Alley cropping: past achievements and future directions. Agrofor Syst 23(2-3):141–155. https://doi.org/10.1007/BF00704912

Lal R (1997) Soil degradative effects of slope and maize monoculture effects on a tropical Alfisols in Western Nigeria and soil physical properties. Land Degrad Dev 8:325–342

Lal R (1999) Soil management and restoration for C sequestration to mitigate the greenhouse effect. Prog Environ Sci 1:307–326

Lal R (2006) Enhancing crop yields in the developing countries through restoration of soil organic carbon pool in agricultural lands. Land Degrad Dev 17:197–209

Malo DD, Schumacher TE, Doolittle JJ (2005) Long term cultivation impacts on selected soil properties in the northern Great Plains. Soil Tillage Res 81:27–29

Motsara MR, Roy RN (2008) Guide to laboratory establishment for plant nutrient analysis, FAO fertilizer and plant nutrition bulletin 19. FAO, Rome

Ong CK, Corlett JE, Singh RP, Black CR (1991) Above and below ground interactions in agroforestry systems. For Ecol Manag 45:45–57

Owoade FM, Adiku SGK, Atkinson CJ, Kolawole GO, MacCarthy DS, Narh S (2017) Residue retention practices for carbon sequestration in some Ghanaian soils: a survey of willingness towards climate change mitigation. In: Proceedings of the international conference on climate change and sustainable development in Africa (ICCCSDA): climate change and sustainable development: strengthening Africa's adaptive capacity. University of Energy and Natural Resources, Sunyani, 25–28 July 2017

Reynolds TW, Waddington SR, Anderson CL, Chew A, True Z, Cullen A (2015) Environmental impacts and constraints associated with the production of major food crops in Sub-Saharan Africa and South Asia. Food Sec 7:795–822

Sanchez P, Denning G, Nziguheba G (2009) The African green revolution moves forward. Food Sec 1:37–44

Shu X, Zhu A, Zhang J, Yang W, Xin X, Zhang X (2015) Changes in soil organic carbon and aggregate stability after conversion to conservation tillage for seven years in the Hung-Huai-Hai Plain of China. J Integr Agric 14:1202–1211

Soil Research Institute (1999) Soil map of Ghana, FAO/UNESCO (1990). Land suitability section. Soil Research Institute, CSIR, Accra

Waddington SR, Li X, Dixon J, Hyman G, de Vicente MC (2010) Getting the focus right: production constraints for six major food crops in Asian and African farming systems. Food Sec 2:27–48

Yilfru A, Taye B (2011) Effects of land use on soil organic carbon and nitrogen in soil of Bale, South eastern Ethiopia. Trop Subtrop Agro Ecosyst 14(1):229–235

Zerihun T (2017) Raising crop productivity in Africa through intensification. Agronomy 7:22; 30. https://doi.org/10.3390/agronomy7010022

4

Women Participation in Farmer Managed Natural Regeneration for Climate Resilience: Laisamis, Marsabit County, Kenya

Irene Ojuok and Tharcisse Ndayizigiye

Contents

Introduction ....
Findings and Recommendations ...
    To Identify Potential Risks Associated with Climate Change and Land Degradation
    Facing the Communities in Laisamis (Mention the Key Threats, Who Is Most at
    Risk, Why) .
    Assess the Opportunities FMNR Has in Increasing the Resilience of the Communities to these
    Risks (Define FMNR, Its Key Benefits from an Environmental, Economic,
    Social View, etc
    What Are the Roles of Men and Women in Uptake of FMNR
    What Drives Women to Participate in or to Adopt FMNR
    Key Barriers Affecting Women Participation in FMNR
    Recommendations for Enhancing Women Participation in FMNR Uptake .
Limitation of the Study
Conclusion
References

I. Ojuok (✉)
National Technical specialist Environment and Climate Change, World Vision Kenya, Nairobi,
Kenya
e-mail: Irene_ojuok@wvi.org; barrackawino@yahoo.com

T. Ndayizigiye
SMHI/Swedish Meteorological and Hydrological Institute, Nairobi, Kenya
e-mail: Tharcisse.Ndayizigiye@smhi.se

Abstract

Despite the fact that land degradation is both natural and human-induced, it is proven that human activities pose greatest threat and these include unsustainable land management practices such as destruction of natural vegetation, over-cultivation, overgrazing, poor land husbandry, and excessive forest conversion. Other than reduced productivity, land degradation also leads to socioeconomic problems such as food insecurity, insufficient water, and regular loss of livestock which exacerbate poverty, conflicts, and gender inequalities that negatively impact mostly women and children especially the rural population. Increased efforts by governments, donors, and partners toward reversing land degradation through community-led, innovative, and effective approaches therefore remain to be crucial today than never before!

Farmer-managed natural regeneration (FMNR) is a proven sustainable land management technology to restore degraded wasteland and improve depleted farmland. This approach has been tested across Africa with high success rates. In spite of the huge local, regional, and global efforts plus investments put on promoting FMNR across different landscapes among vulnerable communities for climate resilience, the implementation of such projects has not been as successful as intended due to slow women uptake and participation in the approach. In order of ensuring women who are mostly at highest risk to impacts of climate change enjoy the multiple benefits that come along with FMNR, the success rate for uptake of FMNR especially among women need to be enhanced.

This chapter seeks to explore drivers and barriers of women participation in uptake of FMNR for climate resilience. Findings will be shared from a 3-year project dubbed Integrated Management of Natural Resources for Resilience in ASALs and a Food and Nutrition project both in Laisamis, Marsabit County, Kenya. The program interventions on natural resource management for livelihoods seek to integrate gender and conflict prevention and prioritize sustainable, market-based solutions to address the persistent challenges. The chapter discusses findings, successes, and lessons learned from the actions and the requirement to position women as vulnerable groups at the center of initiatives designed to address the climate change crisis. The outcome of this chapter will enhance gender-responsive FMNR programing through awareness creation, effective organization/project designs, strategies, and plans together with advocacy and policy influence. Limitations of the study and main recommendations for future programing in similar contexts are also shared.

Keywords

Gender · Mainstreaming · Women · Participation · FMNR Uptake · Climate resilience · Marsabit · Kenya

Introduction

Climate change is so far viewed as one of the greatest challenges facing humanity manifested in the form of variation in amount and distribution of precipitation, ocean salinity, wind patterns, and aspects of extreme weather leading to droughts and flooding, among others. These changes threaten community livelihoods, economy, ecosystems, and social cohesion. Africa is particularly viewed to bear the brunt of the climate change threats mainly due to its poor economic development and low institutional capacity. Vulnerable communities especially women and children within the continent are facing the highest pressure following conspicuous threats like decline in crop production, livestock deaths due to droughts, malnutrition, resource-based conflicts, and migration (Fig. 1).

Globally, women are underrepresented in natural resource management (NRM) and peace building, and they are also disproportionately affected by poor NRM because of gendered power relations that deny women access to resources such as land. Women especially in the rural areas are primary providers of water, food, and energy at the household level and therefore bear heaviest impact of poor resource management. In many of the ASALS areas of Kenya, the frequency and magnitude of droughts, floods, and famine have increased significantly in the recent past. This has impacted nega-

Fig. 1 Women in search of firewood, Marsabit, Kenya, February 2019. (Photo: Irene Ojuok, World Vision Kenya)

tively on the pastoralist communities and especially the women whose vulnerability to the impacts of climate change is necessitated by their roles at household level as caregivers. In Kenya, tens of thousands of hectares of farmlands have become so degraded that they no longer produce adequate or regular crops or pasture for livestock. Historically, there has been a male bias in development program research, planning, and implementation activities, which ignore women's role in dryland development and the challenges that they face. Moreover, women generally do not participate in the decision-making processes in the community. Increasingly, women and youth are usually the most marginalized groups often suffering first in situations of severe food and water shortage. They carry the largest burden of searching for fuelwood, pasture, cooking food, and gathering wild fruits, among others. Poor management of natural resources can lead to overuse and degradation, desertification, deforestation, soil erosion, declining water tables, and other effects that can threaten livelihoods and peace. Sometimes, this leads to conflict at household or even between communities. Forest goods and services are largely public in nature and therefore depend to a large extent on public funding. However, prioritization of public investment and incentives to the private sector for forestry development has been low partly as a result of low valuation of forestry goods and services leading to very slow growth of the sector. This has sometimes accounted for the increased utilization of the forest resources beyond its capacity to sustainably be replenished especially due to the fact that increased population and urbanization are largely dependent on these resources. In forest management, women's traditional roles (e.g., collecting water, growing food, etc.) are particularly crucial in drylands in terms of natural resource management and food security. Men have usually been responsible for decision-making and planning of farming activities especially where land is productive with high economic returns, but with the increasing land degradation, they leave the degraded areas to look for jobs in urban areas, leaving women to assume new roles and responsibilities on the farm. In such a changing context, it is fundamental to be aware of the obstacles hindering full participation of disadvantaged groups, including women in the adverse climate change conditions (Figs. 2 and 3).

Kenya has cited climate change as one of the most serious challenges affecting achievement of its development goals as described under Vision 2030. Toward this end, in 2010, Kenya developed a National Climate Change Response Strategy (NCCRS) followed by the National Climate Change Action Plan (NCCAP) in 2012, which recognized the importance of climate change impacts on the country's development and identifies significant adaptation measures in various sectors of the economy. However, there has been lack of gender-sensitive targets and indicators purposefully set to track progress. There is also need for capacity-building with respect to climate change and its implications for achieving Kenya's gender and development targets. Climate risk management efforts could equally be strengthened by improving understanding of the implications of climate change for Kenyan women and other vulnerable groups, deepening analysis of how to develop and implement adaptation measures that minimize adverse impacts on these groups, and integrating this knowledge into policy and program implementation. Creation of this understanding will likely need to be supported by extensive research and more awareness-raising with policy-makers and experts in this field.

Fig. 2 Displaced family moving to new place for settlement due to drought extreme in Marsabit County. (Photo: Irene Ojuok)

Fig. 3 Women practising FMNR in Laisamis, Marsabit County. (Photo: Irene Ojuok)

Farmer-managed natural regeneration is a low-cost, sustainable land restoration technique used to combat poverty and hunger among poor subsistence farmers in developing countries by increasing food and timber production and resilience to climate extremes. It involves the systematic regeneration and management of trees and shrubs from tree stumps, roots, and seeds. World Vision has successfully implemented FMNR as a sustainable forest management approach among communities with tremendous uptake by communities especially in ASAL areas due to the sustainable benefits derived from the tree management systems. Among the tree based value chains are like sale of fuelwood, pasture bulking, and beekeeping which are some of the drivers identified that have motivated farmers to increase restoration efforts due to increased returns from wood and non-wood products. The main beneficiaries of this approach are those who depend more on tree resources: farmers, herders, and particularly women and children who harvest wood and non-timber forest products, which are based on documented economic values of FMNR.

There is, however, a need for in-depth analysis and dissemination of findings on the role of women in natural resource management with focus on drivers of women participation on FMNR for increased climate change resilience in ASAL counties of Kenya. In every society, there are marginalized groups, often including women and children, people with disabilities, and minorities. External parties can play a critical role in opening the eyes of all to the benefits of inclusion and in facilitating positive change sensitively hence need for this chapter in the climate adaptation hand book.

The case study for this chapter draws on World Vision that has wealth of experience in implementing FMNR approach which has been successful in different contexts in Kenya over the last 8 years including in semiarid areas. Between 2018 and 2021, World Vision Kenya (WVK) in partnership with Northern Rangeland Trust (NRT) and Stockholm Environment Institute (SEI) will be implementing a project dubbed Integrated Management of Natural Resources for Resilience in ASALs (IMARA). The goal of the project is toward increasing resilience of about 35,000 marginalized households to climate change-related shocks through diversified livelihoods and improved natural resource management and use in the ASAL Counties of Isiolo, Laikipia, Marsabit, and Samburu through funding from SIDA. The key objectives are as follows: (1) secure livelihoods and strengthened market systems (including for women and youth) that support sustained management of natural resources; (2) sustainable management and rehabilitation of land, forest, and water sources for strengthened ecosystem services; and (3) strengthened governance systems and structures for sustainable NRM at community, county, and national levels. As part of baseline and program intervention mapping, a gender and social inclusion assessment were conducted in the target program area to establish key sociocultural, economic, political, and technological issues that affect participation of women and youth. This is in line with the Swedish Development Cooperation Strategy for Kenya (2016–2020) which includes a focus on "A better environment, limited climate impact and greater resilience to environmental impacts, climate change and natural disasters."

To complement these objectives, the proposed case study will be limited to identifying the key factors that drive or bar women participation in adoption of FMNR approach for climate resilience in Marsabit County. The findings will be used to develop strategies to effectively and meaningfully engage women in all stages of program implementation cycle. The study findings will strive to enhance women, girls, boys, and men in the community participate in a process which allows them to express their needs and to decide their own future with a view to their empowerment in sustainable NRM, livelihoods, and governance domains for increased resilience to impacts of climate change.

The study opted social research using qualitative survey approach. The main method was through interviews using focused group discussions (FGD) and key informant interviews (KII). In this process, KII guide and FGD guides were the main tools adopted for use during the research. The sampling was purposive since it was intended to get information from targeted audience with knowledge on natural resource management and FMNR approach. The interviews entailed for ten sets of respondents (three FGD comprising women only, men only, and mixed group of men and women of different age groups, KIIs including two government officers from environment and agriculture departments, woman FMNR champion, youthful male and female practicing FMNR, World Vision staff implementing the project in Marsabit, and a local chief). An interpreter from the local community supported translations during the interviews due to language barriers in some instances. Summary tables based on the study themes were employed for analysis using the QDA Miner Lite version 2.0.6. The analysis results were presented in terms of tables, graphs, and pie charts. Secondary data sources sourced from document review using document review matrix guide aligned with the key study objectives. Observations were conducted during field travels while doing data collection in the villages of Marsabit. Photography was equally used to obtain general status of the landscape encompassing key features, e.g., vegetation and habitat, migrations, human settlement, and human activities, e.g., trade, livestock rearing, land restoration efforts, etc. (Fig. 4).

Findings and Recommendations

To Identify Potential Risks Associated with Climate Change and Land Degradation Facing the Communities in Laisamis (Mention the Key Threats, Who Is Most at Risk, Why)

Overall, *extreme climatic conditions* linked to climate change mostly mentioned by the respondents include severe and prolonged drought, increased flood occurrences (El Niño), extreme heat caused by high temperatures, low and irregular rainfall

Fig. 4 FMNR benefits – gums and resins, flowers, and seeds used as livestock feed critical in drought season. (Photo: Irene Ojuok)

patterns, and disease outbreaks, e.g., malaria and diarrhea. In the month of April 2019 (Kenya 2019), seven people were confirmed dead following the outbreak of **kala-azar** vector (sand fly) disease in **Marsabit County mainly** in Laisamis sub-county. This insect is most active in humid environments during the warmer months and at night, from dusk to dawn. According to the World Health Organization (WHO) 14 March 2019 report, this disease which mainly affects the poorest people on Earth is linked to environmental changes such as deforestation, building of dams, irrigation schemes, and urbanization. Environmental management hence cited important in reducing or interrupting transmission of this disease (Fig. 5).

Intensified **pressure on natural resources** is attributed to water scarcity, limited pasture, increased deforestation, limited access to firewood, soil and wind erosion, and along the river beds overharvesting of sand. Risks associated with climate change have also led to **loss of livelihoods and food insecurity** leading to death or poor health of livestock, low crop yields, increased charcoal burning for sale, and loss of wild fruits and seeds/fodder usually consumed by human and livestock during severe drought. Hunger and malnutrition manifest itself in these cases. Some of the social challenges include **displacements/migration**; **conflict over natural resources**; **increased crime**, e.g., cattle raiding; **school dropouts**; **family breakdown**; and **poverty**. One of the respondents emphasized that during drought "only one out of five children attends schools" hence higher levels of illiteracy in the region. In a men FGD, one respondent expressed threats associated *with morans marrying their women* since the younger men can offer better living conditions than the older ones.

91% of the respondents reported that **women and children are the most at risk** in these circumstances. The reproductive roles undertaken solely by women were emphasized to increase their vulnerability in negative climate change impacts. For instance, when men migrate in search of pasture, women take up full responsibility

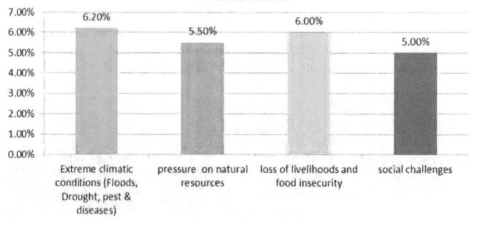

Fig. 5 Analysis of risks associated with climate change

for households; conflicts due to pressure on natural resources leave women most affected because of their weak levels of defense, and hunger caused by drought deteriorates health of pregnant and lactating mothers including children under five. The men-only FGD confirmed that a *woman had been killed in the process of protecting her children* when an enemy attacked their household and the man had migrated with large herds in search of pasture. Limitations on decision-making and utilization of valuable resources that can help support them while in need, e.g., livestock, land, and large-scale sale of tree value chains, are mainly men-dominated, and they shared justifications to this. This emphasizes the critical need of enhancing women participation in sustainable natural resource management while exploring the potentials FMNR has in increasing their resilience to climate change (Fig. 6).

Assess the Opportunities FMNR Has in Increasing the Resilience of the Communities to these Risks (Define FMNR, Its Key Benefits from an Environmental, Economic, Social View, etc.)

The FGDs and KIIs were sought to explore FMNR benefits in resilience-building among the communities (Fig. 7). Adaptation and mitigation opportunities came out strongly with environmental conservation benefits leading, followed by enhanced availability of pasture for livestock and increased food security which aside from addressing the domestic household needs also offered better economic opportunities. Respondents confirmed that FMNR is an **adaptable approach to land restoration** especially in an ASAL context since it is easy to apply and inclusive (done by men, women, people living with disabilities, and children) and has high success rates,

Fig. 6 Analysis of the most affected gender in climate change risk scenarios

PERSONS AT RISK.

9%

91%

- women and children at risk
- men at risk

Fig. 7 Opportunities FMNR provide for enhanced resilience

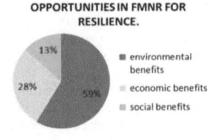

OPPORTUNITIES IN FMNR FOR RESILIENCE.

13%

28%

59%

- environmental benefits
- economic benefits
- social benefits

community bylaws, and regulation control in tree harvesting, thus reducing pressure caused by overharvesting of trees. Invasion of *Prosopis juliflora* (*mathenge*) in Marsabit is a menace, but FMNR management practices applied to this species have enabled better management. Increased **access to high-value indigenous trees**, e.g., wild fruits and Acacia pods, was cited by the pastoralists especially women as a critical feed for human and livestock especially during severe drought and for medicinal purposes. Often, women are the repository of extensive botanical knowledge since they know the traditional uses of many species. One woman says "Such trees are a mystery to us, without which we perish. Acacia tree pods and flowers can feed livestock for 10 months in scarcity of grass." **Less migration of people and livestock** is experienced where FMNR is practiced due to availability of pasture/fodder throughout the year. This reduced the burden of women being household heads during prolonged drought. They also mentioned the link between **improved health and productivity of their livestock** which ultimately addresses food security and economic empowerment since their livestock fetches higher market rates and increased milk and meat production offers nutrition needs as well as excess for markets. FMNR strengthens knowledge in tree-based value chains, thus enhancing **diversification of livelihood opportunities** from wood and non-wood forest products (NWFPs), e.g., trade in gums and resins, honey, gum arabic, firewood, acacia pods, wild fruits, etc. Respondents also brought out the FMNR social aspects in which cultural and religious attachments to specific species of trees enhanced their protection and management, e.g., there are species used during marriage, circumcision, and sacred events.

What Are the Roles of Men and Women in Uptake of FMNR

The objective explored the different roles played by men and women in the uptake of FMNR practice. This was based on the domestic chores undertaken by men and women in the family and community setup. It was evident that women are directly responsible for **care and management of trees** due to many derived benefits they accrue from the trees that enable them run their families. In all the mentioned roles of women, protection and use of the tree resources were most significant. Men, on the other hand, mainly offer **administrative roles in control of the natural resource**. Out of the nine roles played by men in uptake of the practice, five are mainly in management and control of use of the resources. Both men and women have equally **economic interest** role rates in the practice of FMNR. **Trade** in livestock, gum and resin, seeds, and wild fruit collection is done by both. **Management and utilization** of tree resources especially in large scale is reserved for men. It was interesting to note that **development and enforcement of bylaws or regulations** for environmental protection significantly sit on men (Fig. 8).

Generally, women often manage sheep and goats as they tend to be kept closer to the homestead. Women also tend to be left responsible for the home herd of cattle and camels when men take others on migration. As such, women's roles in livestock and environmental management should not be underestimated, and often their

No.	Men	Women	Both Men and Women
1.	Decision making on management and use of trees	Taking care of trees	Prunning and thinning of trees
2.	Control over the tree resources	Feeding the family and small stock from tree products	Offer education on environmental conservation
3.	Administration of justice especially for tree offenders	Putting up shelter for the family using tree branches	Harvesting of Wood and NWFPs
4.	Only men practise bee keeping	Creating awareness and trainings to other women on FMNR	Trading in wood an NWFPs
5.	Guide ceremonies in which specific trees are used	Reporting tree offences to men/elders	
6.	Educating young boys on importance of taking care of the earth during initiation ceremonies e.g. circumcision	Small scale harvesting of Wood and NWFPs (firewood, acacia pods, wild fruits, herbs, gums and resins)	
7.	Long and short distance grazing for large herds especially during drought	Trading in wood and NWFPs to feed the family	
8.	Large scale Harvesting of trees especially mature/very old ones	Short distance grazing especially for small stock	
9.	Developing bylaws/regulations for tree management and use	Protecting and watering of trees e.g. the wildings	

Fig. 8 Roles of men and women in uptake of FMNR

knowledge on livestock as well as grazing areas, migration routes, and water points is rich. Women and men typically have different objectives for keeping animals, different authorities and responsibilities, and different abilities to access and use new information and improved technologies. These differences may lead them to have different priorities regarding investments in the adoption of new technologies and practices and/or different ideas about how best food and livelihood security can be attained through embracing FMNR.

What Drives Women to Participate in or to Adopt FMNR

From both FGD and KII interviews, FMNR ability to enhance **access to immediate household basic needs**, e.g., firewood, feeds (for human and livestock), building and fencing materials, and beddings, was cited as the most important driver to women participation in uptake of this practice (Fig. 9). Continued deforestation in Laisamis has led women walking long distances in search of firewood; one women reported walking 8 h (to and fro) to collect water and firewood from far off hills. Acacia pods offer nutritious feed for livestock especially during drought when pasture is limited, wild fruits have been a significant source of food in households, both seeds and wild fruits are also sold in the market, and women are mostly involved in this business. Alongside this, **economic returns from sale of products** from the environmental services accrued as a result of FMNR boosts women to participate in savings and loaning schemes, thus enabling them to expand and diversify their businesses/livelihood opportunities and increase their income levels. As they easily access basic immediate need through FMNR, women have more time to explore participating in other interventions like poultry keeping and kitchen gardening for nutrition and markets. One of the respondents noted that "Women

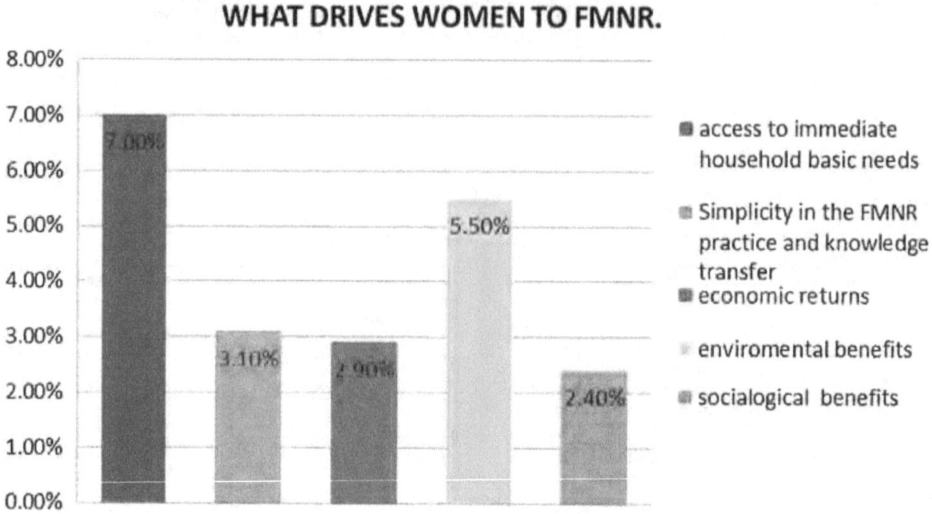

WHAT DRIVES WOMEN TO FMNR.

Fig. 9 Key drivers of women uptake and participation in FMNR

will go to great lengths to extract resources they need to take care of their families including entering dangerous insecure areas such as no-go zones which put them at risk of rape, kidnapping and killing." Improved health of livestock enhances the prices and increases milk production which is valuable to women. **Ease in application and transferring of knowledge** and practice of FMNR including its **social benefits was ranked third** in women drivers in uptake of FMNR. Participants affirmed the low-cost approach would best suit women since they are most vulnerable, and the more expensive an approach, the lesser application they would have of it. Majority of women in the region are illiterate; hence, the simplicity in the practice enhances women participation. **Environmental benefits** from FMNR were equally a factor since severely degraded landscapes are key threats to women because this deprives them access to basic ecosystem services. Marsabit is generally a dry place with high temperatures; FMNR increases tree cover that provides shade required by human and livestock. Women appreciated the importance of shade for resting and relaxing after daily chores. In some instances, goats have equally died due to prolonged exposure to very high temperatures, thus underpinning the importance of trees offering shade. Bare land exposes the community to excessive dust which further than contaminating food and causes eye and respiratory problems. Strong winds in the area have led to destruction of property; hence, trees play an important role as they serve as wind breaks.

Key Barriers Affecting Women Participation in FMNR

The main barrier limiting effective women participation in uptake of FMNR mentioned by all the respondents is **excess workload** on women which affects their concentration and consistency in rollout of FMNR since their attention is

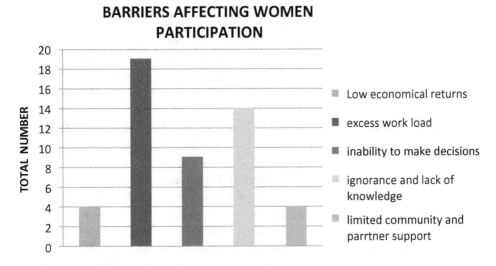

Fig. 10 Barriers affecting women participation in FMNR

diverged to addressing domestic chores and reproductive roles at household (Fig. 10). Severe drought and increased water stress necessitate this challenge overburdening women the more. **Ignorance and lack of knowledge** was cited as threat to uptake of this approach since it is likely to limit transfer of knowledge and skills. Most women in this community are illiterate, for instance, in women-only FGD, 90% of the respondents were only exposed to informal education and could only speak their local language, hence causing communication barriers in new knowledge transfers. Inability to make full decisions over management and utilization of tree resources as well as livestock that have direct linkage to success of FMNR was equally a challenge the women faced as observed by the respondents. Women's subordinate position in society, and their diminished access to information, education, and training, affects their participation in decision-making and public life. This is sometimes attributed to the cultural norms like overstocking by men depicting wealth and pride. Land ownership also rests on men; hence, women have limited power to make decisions on resources on land. Traditional institutions that offer guidance and leadership are mainly comprised of men; in fact, council of elders which is the highest traditional institution is purely composed of men. Women also need the security of knowing that they will have equal decision-making power in how resources and incomes from FMNR work will be used. This is especially important for women-headed households. **Low economic returns** and long-term benefits expected from trees and NWFPs minimize the interest of women in land restoration efforts including FMNR. The respondents acknowledged the many income sources from tree-based value chains, but the low market prices make the venture not worthwhile. For instance, they cited case of gums and resins which cost 200/kg in their locality, yet the same goes for 1000/= in the urban areas. This makes them feel exploited. Lack of organized market systems for the FMNR value chains limits women's return on the investments in the practice, thus derailing their efforts. **Vastness of the area and limited transport** means makes it even more difficult for knowledgeable people to create awareness of the approach due to logistical challenges. **Minimal community and least concerted partner support** to address this challenges derails passion among the existing women champions who strive to promote uptake of FMNR in the community.

Recommendations for Enhancing Women Participation in FMNR Uptake

Intense awareness and improved capacity for women to participate in FMNR was identified as most critical need in influencing their uptake of this approach (Fig. 11). This could be achieved through trainings, exposure to successful sites, and increased access to quality germplasm required by women. Gaps in effective monitoring and evaluation of projects have previously affected success of the projects; the respondents emphasized the need to have **community-led monitoring**, **evaluation**, and **learning model** for FMNR approach which are women-led, inclusive, and responsive. This will ensure women own this approach and integrate it in their daily activities. For instance, women are main beneficiaries of firewood; it's easy for them to include pruning and protecting of trees as they harvest firewood. The respondents also noted low **involvement of community and partners** in environmental conservation initiatives which then becomes a barrier to success of implementing this approach since it requires everyone. **Drought management and increased water access** is key in reducing pressure

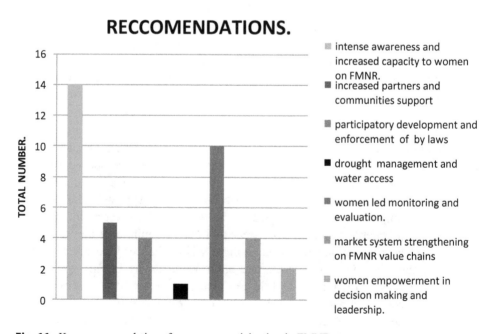

Fig. 11 Key recommendations for women participation in FMNR

women are exposed to in addressing household needs which would then give them time to get fully involved in FMNR activities. This continues to emphasize the need to invest more in adaptation and mitigation efforts to improve community resilience to climate change effects. More efforts are also required to **ensure women meaningfully participate in decision-making and have leadership roles** in natural resource management as they are the most at risk and have the ability to offer best solutions. In circumstances where women and men have equal access to productive resources, control over those resources is usually vested in men. Across all the above objectives, economic returns have been mentioned driver to investments by community in land restoration. While interrogating this further in the recommendations, respondents cited the need to strengthen the **market systems and value chains along FMNR** and other sustainable natural resource management approaches as this will attract more people to invest in the models. There are very many opportunities in FMNR that both address mitigation and adaptation which economic incentives would propel greatly. "Why can't private sector and research actors explore marketing or adding value to acacia pods from acacia tree that is proved to enhance milk production and health in goats and is the main livestock feed of all pastoral communities that take them from one drought season into the rainy seasons? The market for this product is disorganized hence no adequate return on investment, yet this can be big driver for pastoralists to bring back acacia trees on land which provide other multiple benefits" reiterates one of the respondents.

Limitation of the Study

While the case study has established significant findings and recommendations in enhancing women participation in uptake of FMNR for climate resilience, it's important to note that the limited duration in implementation of such findings may not yield to desired outcomes. The study also targeted a narrow geographical extent with small population applying the quantitative and quantitative data collection methods to derive the results; hence, the outcomes need to be understood within this scope. Intended, IMARA project evaluation in 2021 will additionally establish the contributions of these findings to the outcome of women participation in FMNR within Marsabit County. Additionally, it is important to note that in many communities living in ASALs, the status of women and girls is subordinate to that of men and boys. Key assets and resources, such as land, livestock, water, and cash, are generally controlled by older men rather than by women or youth, reflecting the subordinate position of women in society and the cultural limitations placed on their public role. The welfare of women and girls is also directly threatened by environmental problems, which increase the pressures of providing for the household, particularly water and fuelwood collection. Further studies are hence recommended in this field to bridge the gender gaps that continue to challenge women participation climate resilience initiatives or programs.

Conclusion

About 80% of Kenyan women spend about 1–5 h per day per household looking for fuelwood (p. vi), and in some rural locations (particularly in the ASALs), they spend 3–5.25 h a day collecting water (p. 20). Climate disasters can increase women's household responsibilities and cause disproportionate economic losses. During periods of drought, especially in the ASALs, women and children need to walk greater distances to fetch water and fuelwood. The household responsibilities of women can also increase, even more so if the able-bodied male members of a household leave in search of economic opportunities. Greater resource scarcity can also increase the likelihood of women and children being affected by conflict and violence. Women are at greater risk as well during periods of flood, when the occurrence of malaria, cholera, and dysentery – to which pregnant women and children are more vulnerable – can increase, particularly in areas where access to health care is inadequate. Despite women's multiple burdens, they have proved their capacity for effective collective action at the local level and shown that investment in their empowerment generates positive multiplier effects across communities in ways that improve human welfare.

IMARA program interventions on natural resource management for livelihoods seek to integrate gender and conflict prevention and prioritize sustainable, market-based solutions to address the persistent challenges. The efforts will transform the lives of communities and households in the four target counties and revitalize/preserve rangeland assets for the benefit of future generations especially women who are mostly affected by the negative impacts of climate change. According to the County Integrated Development Plans (CIDP) for Marsabit 2017, one of the key social problems in Marsabit County is high gender inequality as women and youth participation in development is low and there are few women and youth involved in leadership and decision-making; traditional and cultural practices are dominant. It is on this basis that the case study sought to investigate the factors affecting women participation in uptake of FMNR for climate resilience with the below summarized conclusion:

- There is evidence of risks associated with climate change linked to climate variability (drought, floods, high temperatures, disease outbreak), increased pressure on natural resources leading to resource-based conflicts, loss of livelihoods intensifying poverty, and social disintegrations causing lack of cohesion among communities. These increase vulnerabilities of communities, and 91% of the respondents noted that women and children as the most affected.
- Environmental, economic, and social benefits proved greatest opportunities FMNR has in increasing resilience of farmers to impacts of climate change.
- Women were cited to be more involved in the actual restoration and exploitation of the tree resources due to reproductive roles which are heavily dependent on sustainable landscapes. On the other hand, men are mainly into overall administration and management of the resources. This means women invest in what they have limited gains especially when there are huge returns.

- Main drivers to women participation in FMNR included access to immediate practical needs, environmental benefits, simplicity and adaptability of the approach, and economic and social benefits. These are in order of priority.
- The greatest barrier threatening women participation in FMNR is excess work load on women in the household following limited efforts to address the burden of women especially in ASAL communities. Ignorance attributed to high illiteracy levels, low decision-making powers, low economic returns, and inadequate partner and community support limits women participation.
- Recommendations, intense gender-responsive FMNR campaigns, women-led monitoring, evaluation and learning, partner and community support, market systems strengthening for FMNR value chains, participatory development and enforcement of bylaws, women empowerment in leadership and decision-making, and finally drought management and increased water access (Figs. 12, 13, 14, 15, 16, and 17).

Fig. 12 FGD participants in sampled sessions. (Photo: Irene Ojuok)

Fig. 13 Typical house in Laisamis constructed by woman and children enjoying on tree. (Photo: Irene Ojuok)

Fig. 14 Bare land that is severely degraded in Laisamis and another on FMNR practice. (Photo: Irene Ojuok)

Fig. 15 Alternative livelihoods women engage in where FMNR is working with their support. (Photo: Irene Ojuok)

Fig. 16 On the left the "Tony Rinaudo" tree – as the community call it in July 2019. On the right, the same tree in January 2020, Iltepes women group FMNR site in Korr, Marsabit County. (Photo by Wesley Koskei WVK Communications Officer)

Fig. 17 Tony Rinaudo right livelihood laureate and founder of FMNR on the right with communities and World Vision staff in the field demonstrating FMNR practice, on the left is FMNR integrated with reseeding in Laisamis, Marsabit County. (Photo by Wesley Koskei WVK Communications officer)

Acknowledgments Special appreciation to SIDA for sponsoring this program through the International Training Program for Climate change Adaptation and Mitigation with support from World Vision Kenya and World Vision Australia FMNR Hub.

References

Coppock D et al (2013) Cross-border interaction spurs innovation and Hope among pastoral and agro-pastoral women of Ethiopia and Kenya. Rangelands 35(6):22–28

https://www.google.com/url?sa=t&rct=j&q=&esrc=s&source=web&cd=8&cad=rja&uact=8& ved=2ahUKEwjkhaiszrniAhWNERQKHcjXBHYQFjAHegQIBBAC&url=http%3A%2F% 2Fwww.icpac.net%2Fimages%2Fpublications%2Fadministrative%2FICPAC_MAJOR_ ACHIEVEMENTS_2018.pdf&usg=AOvVaw3xGTmRPrVJUfb4lFKd4nL4

https://www.the-star.co.ke/.../2019-04-24-seven-people-die-of-kala-azar-in-marsabitfmnrhub.com. au/wp-content/uploads/2013/09/Rinaudo-2007-Development-of-FMNR.pdf

https://www.unenvironment.org/news-and-stories/press-release/livelihoods-development-critical- post-conflict-peacebuilding

ICPAC and the World Food Programme (WFP) Atlas (2017) Atlas of Climate Risk and Food Security in the Greater Horn of Africa Region

Nicholson SE (2001) Climatic and environmental change in Africa during the last two centuries. Clim Res 17:123–144

Republic of Kenya (ROK) (2000) National gender and development policy. Ministry of Gender, Sports, Culture and Social Services. Retrieved from http://bit.ly/P1CtN1

5

Adaptive Capacity to Mitigate Climate Variability and Food Insecurity of Rural Communities Along River Tana Basin, Kenya

David Karienye and Joseph Macharia

Contents

Introduction
   Impacts of Climate Variability
   Adaptive Capacity to Mitigate Climate Variability Impacts . . . .
Impacts and Adaptation Strategies to Climate Variability in Arid and Semiarid Lands:
A Case of Garissa and Tana River Counties in Kenya
   Rainfall and Temperature Impacts on Food Security
   Community Perception on Climate Variability and Its Impacts
   Adaptations Strategies to Climate Variability in Arid and Semiarid Land . . .
Conclusions . . . .
Recommendations
References

Abstract

Climate variability is one of the leading natural threats and a root cause of food insecurity in the developing world, more so in Africa. It is a major impediment to the accomplishment of the global Sustainable Development Goals (SDGs), Vision 2030 and Big Four agenda in the Kenyan context. The rise in occurrence and brutality of extreme events resulting from variability of climate including prolonged flooding and drought has become more pronounced in the relatively

D. Karienye (✉)
Department of Geography, Garissa University, Garissa, Kenya

J. Macharia
Department of Geography, Kenyatta University, Nairobi, Kenya

drier areas. This chapter presents a synthesis about rural communities in Garissa and Tana River Counties, Kenya. The key environmental conditions that face the rural communities in the two counties are prolonged drought and recurrent flooding events. The two conditions have resulted in various challenges facing the communities in these regions through low agricultural production (food and pastures), poor infrastructure, human displacement, and the resultant extreme poverty, overall food insecurity, and tough livelihoods. The problems have been exacerbated by lack of capacity by most of the community members to cushion themselves against these impacts. However, as the conditions continue to manifest themselves, the community members have also identified adaptive mechanisms that are best suited in the region including planting drought-resistant crop varieties, diversifying their livelihoods, embrace sustainable land use, and made efforts to plant trees. We, therefore, conclude that integrated information sharing including early warning alongside affordable and appropriate technologies and crop insurance could be an entry point in cushioning the local communities in the arid and semiarid lands (ASALs) against the extreme weather conditions experienced in the region.

Keywords

Adaptive capacity · Africa · Climate variability · Food insecurity · Mitigation · Rural livelihoods

Introduction

Climate variability has been on the rise due to increased global atmospheric greenhouse gas emissions (GHG) comprising mainly of nitrous oxide, carbon dioxide, and methane (IPCC 2014). Carbon dioxide is the key GHG, while as much as methane and nitrous oxide are emitted in trivial quantities in reference to carbon dioxide, they play a significant role in global warming and their associated global effects. For example, N_2O, a potent gas with a high potential to deplete ozone layer, is over 265 more powerful while CH_4 is 28 more powerful in their global warming potential relative to carbon dioxide, over 100 years' time limit (IPCC 2014). These three main GHGs accounts more than 80% to the present global radiative imposing to enhanced global warming and consequently climate variability and its negative resultant effects (Myhre et al. 2013).

Climate variability characterizes one of the extreme economic, environmental, and social intimidations facing the earth presently (Nnadi et al. 2019). In emerging countries, climate variability has a substantial influence on the livelihoods and living situations of the rural communities. Sub-Saharan Africa (SSA) is a "vulnerability hot spot" of climate variability influences (Asfaw et al. 2018). SSA challenges on adaptation will raise considerably, even if the global emission gap is maintained lower than 2 °C due to limited adaptive capacity. The IPCC's Fourth Assessment (AR4) demonstrated that Africa's vulnerability to the effects of climate variability is relatively high due to low adaptive capacity and over-reliance on natural systems for their livelihoods (Mpandeli et al. 2019). Extreme occurrence of droughts is likely to become more rampant and severe in Africa (Schellnhuber et al. 2012). Consequently, climate variability is negatively affecting agricultural production, particularly in SSA where most countries rely heavily on rainfed agriculture as the mainstay of their economies (Abdul-Razak and Kruse 2017). Climate variability related to biophysical stressors is expected to worsen the existing vulnerabilities by dipping the crop yields (González-Orozco et al. 2020).

It is postulated that warming more than 3 °C worldwide will see almost all of the current crops such as maize, sorghum, and millet-cultivated regions in Africa becoming unfeasible for present cultivars. Water unavailability, lower feed quality which is inaccessible, and effects of disease and heat stress will negatively affect production in the livestock sector (Schaeffer et al. 2013). According to Huq et al. (2004), climate variability has a direct impact on how humans manage natural resources and which results in food insecurity. The risks associated with climate variability threaten the capacity of livelihoods to meet basic needs, such as food and water. These effects will be more intense in the arid and semiarid lands (ASALs) where the resources are already limited, vulnerable, and could, therefore, suffer the most.

To mitigate climate variability, community adaptive capacity must be pursued. According to Levina and Tirpak (2006), the term adaptive capacity has been defined differently by different authors. Different authors have explained the concept of adaptive capacity to simply mean the capacity of a natural system to positively respond to the impacts of climate variability. Policymaker also use the term adaptive capacity to refer to the ability of individual communities to respond and adjust their way of life based on the effects of climate variability and lead to adaptation. Therefore, whenever we use adaptive capacity, society and communities must come up with coping strategies especially when dealing with impacts of climate variability in order to minimize its adverse effects.

Communities living in SSA are facing climate variability in a very tough way due to their lack of capacity to respond. The influences include increasing temperatures, more inconsistent rainfall, and increasing incidence of floods and droughts (CARE and ALP 2013). These impacts have severe consequences especially among the rural poor whose livelihoods are directly pegged on the very vulnerable environment. These communities heavily depend on land resources for agricultural production and therefore the impacts of climate variability have a direct impact on their livelihoods. Crop yields will decline transversely in the landmass as ideal growing temperatures are surpassed and growing periods reduced. The areas and timing of cropping activities that were previously suitable for certain crop are anticipated to shift as home-grown climates varies.

In Kenya, the adverse effects of climate variability have also been witnessed particularly in the ASALs which forms ~80% of Kenyan land mass (582,646 km^2) (Macharia et al. 2020). The main effects of climate variability in Kenya have been demonstrated by prolonged and frequent droughts, floods, resurgence of diseases, pests, and environmental disasters. As a result, agricultural productivity is significantly reduced, resulting to increased food insecurity and threatened livelihoods which in most instances leads to human conflicts over scarce land and water resources (Enya et al. 2013). For instance, the *La* Nina occurring between 1999 and 2001 in SSA was the most prolonged and most severe ever, causing devastating effects especially on human livelihoods. The drought affected over four million people due to crop failure and the resultant reduced yields. Droughts have caused starvation, loss of life, and degradation of the environment as a result of deforestation.

Variability in climate poses major threats to environmental sustenance, commercial, and sustainable development in rural areas of the arid regions of Kenya. In particular, the ASAL region of Garissa and Tana River County has been experiencing severe prolonged drought and flooding despite having River Tana traversing the region. This has led to loss of vegetation cover, drying of water catchment areas, rivers, and seasonal streams. This is then followed by heavy lack of pastures and shortage of drinking water resulting to livestock deaths. Recently, in short rains of 2018/2019, Garissa and Tana River counties experienced floods which caused severe havoc resulting to over 50 fatalities, over 15,000 people displaced, and thousands of livestock killed as a result of bursting of River Tana's banks. In addition, extreme weather events such as flooding has spoiled or destroyed transport and communication networks and affected other nonagricultural portions of the food system badly. This has led the communities to seek alternative ways of meeting their livelihoods such as charcoal burning hence environmental degradation resulting to double tragedy from the loss of their only source of livelihood and land degradation. Against this backdrop, this study aimed to close the gap by identifying possible adaptive capacity of the vulnerable communities in the region for the purpose of coping mechanism. This study was conceived to explore the existing adaptive capacity which is sustainable and viable and which the communities would easily embrace to act as adoptive buffer towards the impacts of climate variability in Garissa and Tana River Counties.

Impacts of Climate Variability

Key among the impacts of climate variability includes the following.

(a) **Drought**

Drought is a major threat globally and more so in Africa due to low adaptation capacity resulting from limited resources. Drought results in decreased moisture emanating from inadequate and erratic rainfall and high extreme temperatures. As observed by Keya (1997), moisture storage is largely dependent on rainfall received prior to the onset of drought conditions and the permeability of the soil (micro edaphic conditions). Drought has led to loss of pasture for the livestock as well as wildlife, vegetation loss, and food insecurity. This threatens the source of livelihoods of local communities in the arid lands.

(b) **Loss of biodiversity**

Ecosystem varieties will hypothetically change rapidly as heating increases with reduced precipitation, and will result to biodiversity loss. Some species may be impotent to adapt to the varying climatic conditions (Schaeffer et al. 2013).

High temperatures and lack of precipitation affects distribution and abundance of fauna and flora species. Substantial shifts in climatic situations could result to loss of some standing biomes and the general aesthetic appeal of our environment (Williams et al. 2007). In some instances, the changes in climate may favor the growth of invasive plant species hindering the alien species such as *Prosopis Juliflora* "Mathenge" a common plant in the northern Kenya.

(c) **Food insecurity**

Extreme temperatures above the ideal may have harmful consequences on crop productivity (Wheeler et al. 2000). These changes have a significant effect on the facets of food security since they negatively affect food availability, access, and utilization resulting to unstable and unreliable food systems. Kenya may experience reduced yields with the changing climate (Herrero et al. 2010). Crop yields are drastically reducing in SSA as the optimal temperature increases altering cropping and seasonal calendars (Schaeffer et al. 2013). In Africa and to a large extent, the ASALs region of East African such as Sudan, Ethiopia, and Kenya have in the past experienced hostile climate change. This has hampered crop production leading to acute shortage of food, pastures, and fibers, hence food insecurity. According to Lobell et al. (2011), yields are likely to diminish by ~1% daily for maize crops if such high-temperature regimes are consistent similar with other crops such as cotton and soybeans (Schlenker and Roberts 2009). Similarly, livestock production will be severely affected through quality feed and water availability (Schaeffer et al. 2013).

(d) **Human health and diseases outbreak**

Water availability and increased rates of disease outbreak are transformed by climate change (Schaeffer et al. 2013). The impacts of climate variability will be felt through increased infectious diseases which are relatively high in SSA. Extreme weather events may lead to illness and mortality. The level of malfunction may also be on the rise up to between 35% and 80% due to a rise of between 1.2 °C and 1.9 °C (Lloyd et al. 2011). As reported by Patz et al. (2008), flooding results to disease outbreaks including diarrhea, cholera, trachoma, and conjunctivitis. Other diseases like malaria may shift and be felt to areas where they were not felt before due to changes in temperature suitability responsible for pathogen growth.

(e) **Water resources**

Change in the hydrological sequence due to climate variability has a direct impact on water timing and circulation (Goulden et al. 2009). With most of the countries in SSA facing challenges with provision and supply of clean usable water, climate change will exacerbate this and lead to more water shortages in the coming years (Schellnhuber et al. 2012). The resultant effect will be increased disease outbreaks due to poor sanitation, low agricultural production, food insecurity, and general influence on livelihoods. Rise in temperature due to global warming would lead to a complex rate of evapotranspiration leading to increased loss from water bodies (Ogolla et al. 1997).

(f) **Land degradation**

Population increase combined climate variability impedes good resource management leading to environmental degradation (UNEP 2002b). Climate variability is slowly encroaching and engulfing countries thus rendering their land unproductive due to variations in weather patterns and global warming. As human population grows further, the natural distribution of vegetation on earth will be altered. This leads to opening new land for agriculture and cultivation of marginal areas (UNEP 2002a). This has led to loss of natural habitats, reducing vegetation cover and exposing soils to wind and water erosion in many parts of Africa. Soil erosion has increased the rate of siltation in dams and rivers and at the same time reducing the productivity of the land.

Adaptive Capacity to Mitigate Climate Variability Impacts

Adaptive capacity calls for strategies to help the communities to adapt to these extreme events such as drought and flooding. Adaptation simply means adjustments made in the existing systems as a response mechanism toward countering the effects of climate variability by the communities and individuals involved. These adjustments are mainly meant to act as a buffer and to assure proper exploitation of the new opportunities that minimize harm and as they present themselves. Therefore understanding the adaptive capacity by farmers is crucial to effective adaptation planning since it assures continuous production crucial to effective planning and guarantees human survival (Chepkoech et al. 2020). With the projected increase in global temperature, likely to result to increase in global warming, it's thus inevitable for individuals and communities to find adaptive ways which guarantees their survival. Adaptive measures toward climate change are no longer regarded as second measured but should be taken as primary consideration especially by farmers. However, the adaptation capacity in most African countries is low mainly due to lack of capacity to invest in the recent technologies which have been studied and found to promote better survival and livelihoods. Majority of agriculture in SSA is rainfed with only a very small percentage of farmers with a capacity to carry out irrigation which makes it difficult to predict due to climate variability. Further, the challenges are associated with lack of reliable weather data to inform on policy, and therefore most of the countries lack early warning systems that can be used early enough to caution the governments of possible climate-related calamities.

Impacts and Adaptation Strategies to Climate Variability in Arid and Semiarid Lands: A Case of Garissa and Tana River Counties in Kenya

Rainfall and Temperature Impacts on Food Security

From a data synthesis on annual rainfall and temperature over a period of 20 years for Garissa and Tana River counties indicate that rainfall was characterized with extended dry season occurring between January and February. The long rainy season occurs between March and May (MAM) while prolonged dry season occurs from

Fig. 1 Reduced River Tana flow during the month of October. (Courtesy of D. Karienye)

mid-May to mid-October, while short rainy season begins in mid-October to end of December of each year. There are fewer days of more intense rainfall with the rains often starting late but intense which are described as "very unreliable" (that is seasonal failures are common).

Similarly, for temperatures, the highest temperature amounts were observed between February and March, which coincides with the same time when rainfall is lowest in the study area. Around September–October, the temperatures are also at highest. Temperature increase has an important impact on water availability, thus aggravating drought conditions. Decreases in rainfall have profound repercussions on river flows leading to declining river discharge (Fig. 1). The months that saw an increased rise in temperature also experienced drought. This can be explained by the high evapotranspiration making the vegetation deficient of moisture leading to crop and pasture failure.

However, in trying to escape the droughts, the few well-endowed farmers practiced drip irrigation and greenhouse farming as indicated in Fig. 2. This ensured a reduction in the impact of droughts. This low number of farmers adopting new farming mechanisms, and which is a shift from rainfed agricultural production can be explained by the high cost of the greenhouses' infrastructures. This represents an innovative technology in response to the changing weather patterns though the adoption rates remain relatively low due to high cost.

Community Perception on Climate Variability and Its Impacts

From the interaction with the community, majority of the households were extremely worried about climate variability and identified rainfall to be very unpredictable

Fig. 2 Farmers practicing greenhouses and drip irrigation. (Courtesy of D. Karienye)

Fig. 3 Land degradation through charcoal burning. (Courtesy of D. Karienye)

stating that there exists a consistently prolonged dry period every now and then. Nevertheless, farmers believe that temperatures have already increased and precipitation has declined or is unpredictable (Karienye et al. 2019).

The impact of climate variability has been felt mainly by reduced crop production, extreme cases of flooding, and land degradation as evident by charcoal burning (Fig. 3) and reduced biodiversity. The reduced precipitation coupled with flooding leads to crop failure which destroys the crops that are grown along River Tana. Floods have in the past been responsible for causing disruption in transport systems and displaced residents living in the low-land areas which are prone to flooding (CARE and ALP 2013).

Fig. 4 Community-based agro forestry programs. (Courtesy of J. Macharia)

Adaptations Strategies to Climate Variability in Arid and Semiarid Land

Based on their own experiences and from sharing information among themselves, most of the households in these ASALS of Kenya have identified several adaptive strategies to cushion them against the extreme conditions. The communities preferred livelihood diversification (business, cropping, and livestock) as an alternative livelihood option, sustainable use of the land including conservation agriculture, mulching, building trenches and ditches around the homesteads and watering crops using cans during dry spell. They have also adopted drought-tolerant and early maturing crop species, changing eating behaviors and afforestation (Fig. 4).

Conclusions

From our synthesis, the rainy seasons are no longer predictable thereby prohibiting any farming activities. The impacts of climate variability in the ASALs are mainly through extreme conditions of drought and flooding. The two conditions have resulted to various challenges facing the communities in these regions through low agricultural production (food and pastures), poor infrastructure, population displacement resulting to extreme poverty, overall food insecurity, and tough livelihoods. These challenges are exacerbated further by the inability of the majority of the communities to cushion themselves against the impacts of climate variability and this becomes a cyclic problem year in year out. The better-endowed community members have invested in greenhouses and drip irrigations to ensure continuous supply of food particularly for their domestic consumptions. However, efforts by the local communities have been identified where they have, through experience over time, resulted to planting drought-resistant crop varieties, diversified their livelihoods, embraced sustainable land use, and made efforts to plant trees. It's imperative to note that well-informed, adaptive, and forward-looking decision making is central to adaptive capacity of the host communities. In order for community to respond to expected changes and to participate in adaptive decision-making, they require precise information, knowledge and skills that enable them to actively address climate risks to their livelihoods. Therefore, adaptation energies must aim to ease access to information and the development of the skills and knowledge needed for accurate adaptation targeting. Institutions and agencies responsible for policy formulation should ensure an enabling atmosphere for local adaptation efforts.

Recommendations

In order to embrace the adaptive capacity as long-term practical solutions, the following are recommended:

- Monitoring daily weather patterns and improving scientific understanding of climate.
- The community needs to be trained on affordable and appropriate technologies such as sustainable agriculture.
- Promotion of climate-smart crops farming.
- Promotion of insurance services against the consequences of catastrophic weather events to mitigate against climate variability.
- Provision of early warning systems to the communities.
- There is a need to build community-based capacities in planning, coordination, and implementation of climate change adaptation activities and programs.
- Intensification of tree planting through community-owned nurseries, establish green zones, and invest in reforestation programs.

References

Abdul-Razak M, Kruse S (2017) The adaptive capacity of smallholder farmers to climate change in the Northern Region of Ghana. Clim Risk Manag 17:104–122. https://doi.org/10.1016/j.crm.2017.06.001

Asfaw A, Simane B, Hassen A, Bantider A (2018) Variability and time series trend analysis of rainfall and temperature in northcentral Ethiopia: a case study in Woleka sub-basin. Weather Clim Extrem 19:29–41. https://doi.org/10.1016/j.wace.2017.12.002

CARE, ALP (2013) Climate change vulnerability and adaptive capacity in Garissa County. Care International and Adaptive Learning Program (ALP), p 20. http://www.careclimatechange.org/files/CVCA_Kenya_Report__Final.pdf

Chepkoech W, Mungai NW, Stöber S, Lotze-Campen H (2020) Understanding adaptive capacity of smallholder African indigenous vegetable farmers to climate change in Kenya. Clim Risk Manag 27:100204. https://doi.org/10.1016/j.crm.2019.100204

Enya D, Niversity K, Enya H (2013) National implementing entity – NEMA – Kenya programme, proposal programme title: integrated programme to build resilience to executing entities, October 2013 (Feb)

González-Orozco CE, Porcel M, Alzate Velásquez DF, Orduz-Rodríguez JO (2020) Extreme climate variability weakens a major tropical agricultural hub. Ecol Indic 111:106015. https:// doi.org/10.1016/j.ecolind.2019.106015

Goulden M, Conway D, Persechino A (2009) Adaptation to climate change in international river basins in Africa: a review. Hydrol Sci J 54(5):805–828

Herrero M, Ringler C, van de Steeg J, Thornton P, Zhu T, Bryan E, Omolo A, Koo J, Notenbaert A (2010) Climate variability and climate change and their impacts on the agricultural sector. ILRI report to the World Bank for the project "adaptation to climate change of smallholder agriculture in Kenya". International Livestock Research Institute (ILRI), Nairobi

Huq S, Reid H, Konate M, Rahman A, Sokona Y, Crick F (2004) Mainstreaming adaptation to climate change in least developed countries (LDCs). Clim Pol 4(1):25–43

IPCC (2014) Summary for policymakers. In: Edenhofer O, Pichs-Madruga R, Sokona Y, Farahani E (eds) Climate change 2014: mitigation of climate change. Contribution of working group III to the fifth assessment report of the Intergovernmental Panel on Climate Change. Cambridge University Press, Cambridge, UK

Karienye D, Nduru G, Kamiri H (2019) Socioeconomics determinants of banana farmers perception to climate change in Nyeri County, Kenya. J Arts Humanit 8(8):89–101

Keya GA (1997) Environmental triggers of germination and phenological events in an arid savannah region of northern Kenya. J Arid Environ 37:91–106

Levina E, Tirpak D (2006) Adaptation to climate change: key terms. OECD, Paris

Lloyd SJ, Kovats RS, Chalabi Z (2011) Climate change, crop yields, and under nutrition: development of a model to quantify the impact of climate scenarios on child under nutrition. Environ Health Perspect 119(12):1817–1823

Lobell DB, Schlenker W, Costa-Roberts J (2011) Climate trends and global crop production since 1980. Science 333(6042):616–620. https://doi.org/10.1126/science.1204531

Macharia J, Ngetich F, Shisanya C (2020) Comparison of satellite remote sensing derived precipitation estimates and observed data in Kenya. Agric For Meteorol 284:107875. https://doi.org/ 10.1016/j.agrformet.2019.107875

Mpandeli S, Nhamo L, Moeletsi M, Masupha T, Magidi J, Tshikolomo K, . . . Mabhaudhi T (2019) Assessing climate change and adaptive capacity at local scale using observed and remotely sensed data. Weather Clim Extrem 26:100240. https://doi.org/10.1016/j.wace.2019.100240

Myhre G, Shindell D, Bréon F-M, Collins W, Fuglestvedt J, Huang J, Koch D, Lamarque J-F, Lee D, Mendoza B, Nakajima T, Robock A, Stephens G, Takemura T, Zhang H (2013) Anthropogenic and natural radiative forcing. In: Stocker TF, Qin D, Plattner G-K, Tignor M (eds) Climate change 2013: the physical science basis. Contribution of working group I to the fifth assessment report of Intergovernmental Panel on Climate Change, pp 659–740. https://doi.org/10.1017/ CBO9781107415324.018

Nnadi OI, Liwenga ET, Lyimo JG, Madukwe MC (2019) Impacts of variability and change in rainfall on gender of farmers in Anambra, Southeast Nigeria. Heliyon 5(7):e02085. https://doi. org/10.1016/j.heliyon.2019.e02085

Ogolla J, Abira M, Awour V (eds) (1997) Potentials impacts of climate change in Kenya. Motif Creative Arts Ltd, Nairobi

Patz JA, Olson SH, Uejo CK, Gibbs HK (2008) Disease emergence from global climate and land use change. Med Clin N Am 92:1473–1491

Schaeffer M, Baarsch F, Adams S, de Bruin K, De Marez L, Freitas S, . . . Hare B (2013) Africa's adaptation gap: technical report. Climate change impacts, adaptation challenges and costs for Africa. United Nations Environment Programme, New York. http://www.unep.org/pdf/ AfricaAdapatationGapreport.pdf. Accessed 10 Aug 2016

Schellnhuber HJ, Hare W, Serdeczny O, Adams S, Coumou D, Frieler K, . . . Rocha M (2012) Turn down the heat: why a 4°C warmer world must be avoided (no. INIS-FR-14-0299). Sauvons le Climat (SLC), Paris

Schlenker W, Roberts MJ (2009) Nonlinear temperature effects indicate severe damages to U.S. crop yields under climate change. Proc Natl Acad Sci U S A 106(37):15594–15598. Retrieved from: http://www.pnas.org/content/106/37/15594

UNEP (2002a) Global Environment Outlook 3. Earthscan Publication, London

UNEP (2002b) Vital climate graphics Africa. UNEP Publication, Nairobi

Wheeler T, Craufurd P, Ellis R, Porter J, Vara Prasad P (2000) Temperature variability and the yield of annual crops. Agric Ecosyst Environ 82:159–167

Williams JW, Jackson ST, Kutzbach JE (2007) Projected distributions of novel and disappearing climates by 2100 AD. Proc Natl Acad Sci 104(14):5738–5742

6

Biomass Burning Effects on the Climate over Southern West Africa During the Summer Monsoon

Alima Dajuma, Siélé Silué, Kehinde O. Ogunjobi, Heike Vogel, Evelyne Touré N'Datchoh, Véronique Yoboué, Arona Diedhiou and Bernhard Vogel

Contents

Introduction ....
Description of the Study Area and Methodology

   Methods
Results and Discussion
   Impact on Meteorology
   Impact on the Atmospheric Composition
   Adaptation and Mitigation Strategies .
Conclusions ....
References

A. Dajuma (✉)
Department of Meteorology and Climate Sciences, West African Science Service Centre on Climate Change and Adapted Land Use (WASCAL), Federal University of Technology Akure (FUTA), Ondo State, Nigeria

Laboratoire de Physique de l'Atmosphère et de Mécaniques des Fluides (LAPA-MF), Université Félix Houphouët-Boigny, Abidjan, Côte d'Ivoire
e-mail: alima.dajuma@yahoo.com

S. Silué
Université Peleforo Gon Coulibaly, Korhogo, Côte d'Ivoire
e-mail: sielesil@yahoo.fr

Abstract

Biomass Burning (BB) aerosol has attracted considerable attention due to its detrimental effects on climate through its radiative properties. In Africa, fire patterns are anticorrelated with the southward-northward movement of the inter-tropical convergence zone (ITCZ). Each year between June and September, BB occurs in the southern hemisphere of Africa, and aerosols are carried westward by the African Easterly Jet (AEJ) and advected at an altitude of between 2 and 4 km. Observations made during a field campaign of Dynamics-Aerosol-Chemistry-Cloud Interactions in West Africa (DACCIWA) (Knippertz et al., Bull Am Meteorol Soc 96:1451–1460, 2015) during the West African Monsoon (WAM) of June–July 2016 have revealed large quantities of BB aerosols in the Planetary Boundary Layer (PBL) over southern West Africa (SWA).

This chapter examines the effects of the long-range transport of BB aerosols on the climate over SWA by means of a modeling study, and proposes several adaptation and mitigation strategies for policy makers regarding this phenomenon. A high-resolution regional climate model, known as the Consortium for Small-scale Modelling – Aerosols and Reactive Traces (COSMO-ART) gases, was used to conduct two set of experiments, with and without BB emissions, to quantify their impacts on the SWA atmosphere. Results revealed a reduction in surface shortwave (SW) radiation of up to about 6.5 W m^{-2} and an 11% increase of Cloud Droplets Number Concentration (CDNC) over the SWA domain. Also, an increase of 12.45% in Particulate Matter (PM_{25}) surface concentration was observed in Abidjan (9.75 µg m^{-3}), Accra (10.7 µg m^{-3}), Cotonou (10.7 µg m^{-3}), and Lagos (8 µg m^{-3}), while the carbon monoxide (CO) mixing ratio increased by 90 ppb in Abidjan and Accra due to BB. Moreover, BB aerosols were found to contribute to a

K. O. Ogunjobi
Department of Meteorology and Climate Sciences, West African Science Service Centre on Climate Change and Adapted Land Use (WASCAL), Federal University of Technology Akure (FUTA), Ondo State, Nigeria

Federal University of Technology Akure (FUTA), Ondo State, Nigeria
e-mail: kenog2010@gmail.com

H. Vogel · B. Vogel
Institute of Meteorology and Climate Research, Karlsruhe Institute of Technology (KIT), Karlsruhe, Germany
e-mail: heike.vogel@kit.edu; bernhard.vogel@kit.edu

E. T. N'Datchoh · V. Yoboué
Laboratoire de Physique de l'Atmosphère et de Mécaniques des Fluides (LAPA-MF), Université Félix Houphouët-Boigny, Abidjan, Côte d'Ivoire
e-mail: ndatchoheve@yahoo.fr; yobouev@hotmail.com

A. Diedhiou
Laboratoire de Physique de l'Atmosphère et de Mécaniques des Fluides (LAPA-MF), Université Félix Houphouët-Boigny, Abidjan, Côte d'Ivoire

Université Grenoble Alpes, IRD, Grenoble INP, IGE, Grenoble, France
e-mail: arona.diedhiou@gmail.com

70% increase of organic carbon (OC) below 1 km in the PBL, followed by black carbon (BC) with 24.5%. This work highlights the contribution of the long-range transport of BB pollutants to pollution levels in SWA and their effects on the climate. It focuses on a case study of 3 days (5–7 July 2016). However, more research on a longer time period is necessary to inform decision making properly.

This study emphasizes the need to implement a long-term air quality monitoring system in SWA as a method of climate change mitigation and adaptation.

Keywords

Adaptation strategy · Biomass burning · lowland · Southern West Africa · Modeling

List of Abbreviations

ADF	Abidjan domestic fire
AEJ	African easterly jet
AGL	Altitude above ground level
AOD	Aerosol optical depth
BB	Biomass burning
BC	Black carbon
CDNC	Cloud droplets number concentration
CO	Carbon monoxide
COSMO-ART	Consortium for Small-scale Modelling – Aerosols and Reactive Traces gases
DACCIWA	Dynamics-Aerosol-Chemistry-Cloud Interactions in West Africa
DMS	Dimethyl sulfide
DWD	German weather service
EDGAR HTAP_v2	Emission Database for Global Atmospheric Research Hemispheric Transport of Air Pollution version 2
FRP	Fire radiative power
GFAS	Global Fire Assimilation System
ICON	Icosahedral nonhydrostatic
ITCZ	Intertropical convergence zone
ITD	Intertropical discontinuity
MODIS	Moderate Resolution Imaging Spectroradiometer
MOZART	Model for Ozone and Related Chemical Tracers
NOx	Nitrogen oxide
OC	Organic carbon
PBL	Planetary boundary layer
PM	Particulate matter
STRATOZ	Stratospheric ozone experiment
SW	Shortwave
SWA	Southern West Africa
TOA	Top of atmosphere

TROPOZ	Tropospheric ozone experiment
WAM	West African monsoon
WHO	World Health Organisation
WRF-Chem	Weather Research and Forecasting model coupled with Chemistry

Introduction

Biomass Burning (BB) is one of the major sources of aerosols in Africa after Saharan dust. In the form of submicron accumulation mode BB pollutants are mainly composed of organic carbon (OC) and black carbon (BC) aerosols and carbon monoxide (CO), hydrocarbon, and nitrogen oxide (NO_x), as gaseous pollutants. During the previous decades, an increasing number of studies have investigated the impacts of BB aerosols on radiation, weather, and climate. In general, the African fire pattern is strongly associated with the southward movement of the intertropical convergence zone (ITCZ) (N'Datchoh et al. 2015; Swap et al. 2002). BB occurs as a result of anthropogenic activities, such as land management, livestock grazing, and crop production (Bowman et al. 2011; Stowe et al. 2002; Mbow et al. 2000). BB is an important source of aerosols and trace gases in the atmosphere, with an estimated burned biomass of 3260–10,450 Tg a^{-1} for tropical areas (Barbosa et al. 1999). Hao and Liu (1994) estimated that there was an amount of 2500 Tg a^{-1} (46% of the tropics in total) over the tropical regions of Africa, to which the savanna contributes up to 1600 Tg a^{-1} (30% of the total amount over the tropics).

BB affects the radiative energy budget of the Earth by absorbing and scattering solar radiation and modifies the properties of clouds as they serve as cloud condensation nuclei (CCN) or ice nuclei (IN). The global mean direct radiative effect of BC and OC from BB was quantified as being 0.155 W m^{-2} for their overall effect, with 0.25 W m^{-2} for BC (absorption) and -0.005 to $+0.4$ W m^{-2} for OC (Jiang et al. 2016). BC traps a significant part of solar radiation in the atmosphere, thus reducing the surface incoming radiation as well as low surface sensible and latent heat fluxes (Huang et al. 2016). A modeling study by Walter et al. (2016) investigated the impact of the Canadian forest fires that occurred in July 2010 on SW radiation and temperature using the regional climate model, Consortium for Small-scale Modelling – Aerosols and Reactive Traces gases (COSMO-ART). Downwelling surface SW radiation was found to be reduced by up to 50% below the biomass plume under cloudless conditions due to absorption in dense smoke layers, furthermore leading to a decrease of up to 6 K of 2-m temperature. Surface cooling, as well as a warming in elevated layers, led to an increase in atmospheric stability, which induce a decrease of precipitation.

Thornhill et al. (2018) analyzed 30-year simulation to assess the impact of BB on the regional climate of South America. The simulations found a decrease in the downwelling clear-sky and all-sky SW radiation at the surface by 13.77 W m^{-2} and 7.37 W m^{-2}, respectively. Mean surface temperature was reduced by 0.14 ± 0.24 °C and a mean precipitation decrease of 14.5% was found in the peak region of BB. The authors found that, if BB increases during a particular dry season, the resulting decrease in precipitation may in fact exacerbate drought. Pani et al. (2016) estimated the direct aerosol radiative effects of BB aerosols over northern Indochina from observed ground measurements of the optical properties of such aerosols. They found that the overall mean aerosol radiative forcing was -8 W m^{-2} and -31.4 W m^{-2} at TOA and at the surface, respectively.

Similarly, Huang et al. (2013) investigated the impact of direct radiative forcing of BC on West African Monsoon (WAM) precipitation during the dry season, and concluded that there was a reduction in precipitation in the WAM region due to the radiative effect of BC. They demonstrated that aerosols from the southern African hemisphere significantly reduced convective precipitation, particularly during the boreal cold season, when BB smoke was prevalent. They also highlighted that BB can affect local weather and climate over West Africa. These results suggest that reductions in cloud amount, cloud top height, and surface precipitation are due to a high BC aerosols load in the atmosphere.

During the airborne measurement campaigns of the Stratospheric Ozone Experiment (STRATOZ) of March 1985 and the Tropospheric Ozone Experiment (TROPOZ) of December 1987, Marenco et al. (1990) discovered that large CO concentrations are present over the mid-Atlantic Ocean as well over West Africa, which is evidence of the lofting of BB from Central and Southern Africa. The plumes arising from agricultural burning fumes mix in a 3–4 km deep boundary layer over Africa as a result of convergence, before overriding moister and cooler air and being advected westward over the Atlantic Ocean (Chatfield et al. 1998). In another study, it was shown that BB plumes from Central and Southern Africa are transported each year during the WAM season and carried westward by a jet located at 700 hPa between 2 and 4 km altitude (Barbosa et al. 1999; Mari et al. 2008). In addition, Real et al. (2010) found ozone plumes in the mid- and upper troposphere over the Gulf of Guinea as a result of the long-range transport of BB from Central Africa. Using backward trajectories, Mari et al. (2008) demonstrated that the intrusion of BB into the upper troposphere of the Gulf of Guinea and the northern hemisphere was controlled by the active and break phases of the southern hemisphere's African Easterly Jet (AEJ) during the summer monsoon (July–August 2006).

The number and size distributions of BB aerosols in southern West Africa (SWA) are dominated by the accumulation mode (Haslett et al. 2019). Likewise, according to a modeling study by Menut et al. (2018), BB from Central and Southern Africa has increased the level of air pollution in urban cities, such as Lagos and Abidjan (approximately 150 µg m^{-3} for CO, 10–20 µg m^{-3} for O$_3$, and 5 µg m^{-3} for PM$_{2.5}$). The contribution of BB in PM$_{2.5}$ concentrations from Central Africa increased from ~10% in May to ~52% in July (Deroubaix et al. 2018). Haslett et al. (2019) found a significant aerosol mass concentration in the SWA boundary layer, both over the ocean and over the continental background, with similar chemical properties. They suggested that the upstream (Gulf of Guinea) aerosols originated from Central African BB, and demonstrated that these aerosols affected cloud optical properties but were less sensitive to precipitation.

Although there have been studies, like the above, on the distribution and impact of BB in various regions the impact of BB aerosols from Central and Southern Africa on the meteorology and climate over SWA specifically has not yet been fully investigated.

Research during these past decades has shown that BB aerosols injected into the atmosphere have adverse effects on radiation and climate. Remote pollution from the long-range transport of BB appears to affect the levels of atmospheric aerosol pollution over SWA during the summer monsoon. It originates from Central and Southern Africa as a result of agricultural activities, land management, livestock grazing, and crop production. It is carried westward by the African Easterly Jet (AEJ) toward the tropical Atlantic Ocean, the Gulf of Guinea, and SWA, aloft at the mid-level troposphere (2–4 km). Recently, an intensive field campaign within the framework of Dynamics-Aerosol-Chemistry-Cloud Interactions in West Africa (DACCIWA) in SWA, observed large amounts of the BB as background concentration in the Planetary Boundary Layer (PBL), dominated by OC (Haslett et al. 2019).

In this chapter, then, the changes in atmospheric variables (precipitation, clouds, solar radiation) due to BB aerosols are examined. During a case study of 3 days (5–7 July 2016), two sets of simulations are performed: with and without BB. The differences make it possible to quantify the impacts of BB on the state of the atmosphere. The following research questions are addressed in this work:

- What is the relative contribution of BB aerosols to the atmospheric composition of the SWA region?
- What is the effect of BB on the climate of SWA?
- What are the practical behaviors and actions to be taken in order to reduce the level of pollution over SWA?

Description of the Study Area and Methodology

This section describes the study area, where the data was obtained (section "Study Area"), and the methodology used (section "Methods") to obtain the results.

Study Area

The study domains comprise an outer domain, that is, D1: West Africa, and a nested domain, that is, D2: southern West Africa (SWA), spreading from 3°N to 10°N and 9°W to 4°E (Fig. 1). The latitudinal extent of D1 (18°W–26.6°E; 20°S–24.6°N) covers the Sahel region, as well as the eastern Atlantic Ocean and Central Africa. The wide spatial coverage of D1 allows us to account for the long-range transport of aerosols, such as dust aerosol from the Sahel and the Sahara Desert and BB from Central Africa. The nested domain D2 (9°W–4°E; 3°–10.8°N), located in SWA, is characterized by tropical forest in the South and grassland and savanna in the North. The climate is that of typical tropics zones governed by the North-South shift of the Intertropical Convergence Zone (ITCZ) or the Intertropical discontinuity (ITD) (over the land), which determines the rainy/dry seasons. During the period under

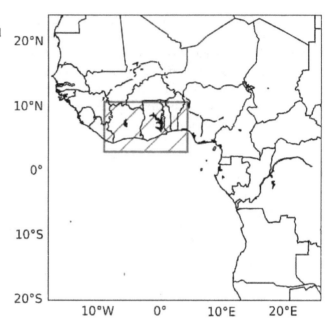

Fig. 1 Model domains D1 (outer domain) and D2 (nested domain) hacked in red. The horizontal resolution in case of D1 is 5 km and in case of D2 is 2.5 km

investigation (July 2016), the WAM was prevalent in the study domain. The WAM is generated by the temperature gradient between the Atlantic Ocean, which is cool (monsoon), and the north (the Sahara Desert), which is warm.

Methods

This section describes the model setup (section "Model Setup") and the biomass burning experiment (section "Biomass Burning Experiment").

Model Setup

Simulations were performed with the regional scale model COSMO-ART (Vogel et al. 2009) at a spatial resolution of 5 km for D1 and nested to 2.5 km for D2 with 50 and 80 levels up to 30 km altitude above ground level (AGL), respectively. Simulations were run for 9 days over D1 (29 June–7 July 2016), and 3 days (5–7 July 2016) were analyzed over D2 as a case study for detailed investigation. COSMO-ART allows for the treatment of the aerosol dynamics, atmospheric chemistry, feedback with radiation, and cloud microphysics (Athanasopoulou et al. 2013; Bangert et al. 2012; Knote et al. 2010; Vogel et al. 2009). A gas flaring emission parameterization of the area of interest in SWA has been developed by Deetz and Vogel (2017), using a combination of remote sensing observations and physically based combustion equations to better reproduce the atmospheric chemistry of the area of interest. The implementation of a 1 D-plume rise model of BB aerosols and gases into COSMO-ART was realized by Walter et al. (2016), which calculates online the injection height (top and bottom limit) of the BB plume. For this study, several emission datasets from different sources have been used. The BB emissions data were obtained from the Global Fire Assimilation System (GFAS) version 1.2 (Kaiser et al. 2012). It is a satellite retrieved dataset of daily fire radiative power (FRP) measurements from the Moderate Resolution Imaging Spectroradiometer (MODIS) Terra and Aqua. Anthropogenic emissions were provided by the Emission Database for Global Atmospheric Research Hemispheric Transport of Air Pollution version 2 (EDGAR HTAP_v2) dataset (Edgar 2010) for 2010 with a 0.1° grid mesh size. In addition, the new gas flaring emission from Deetz and Vogel (2017) was used for this study. Biogenic emissions, sea salt, and mineral dust are calculated online within the model, and mean dimethyl sulfide (DMS) monthly fluxes are prescribed after Lana et al. (2011). The initial and boundary conditions for the meteorology were provided from the global Icosahedral Nonhydrostatic (ICON) model of the German Weather Service (Zängl et al. 2015). The boundary data for gaseous and particulate compounds were derived from the Model for Ozone and Related Chemical Tracers (MOZART) (Emmons et al. 2010). The simulation setup used was similar to the one described in Deetz et al. (2018).

Biomass Burning Experiment

In order to assess the effects of BB over SWA, we carried out two sets of simulations: one with BB emissions (hereafter called Fire) and another one ignoring BB emissions (hereafter called No Fire) for the period under investigation. The experiment with BB includes real-time MODIS observations of fire emissions. Figure 2 presents surface CO emissions for the two case scenarios under investigation.

Due to its long lifetime and its reactivity, CO is used as a surrogate for BB detection. Moderate CO emissions are observed in SWA mostly over the city plumes (Abidjan, Accra, Cotonou, and Lagos) due to local anthropogenic emissions.

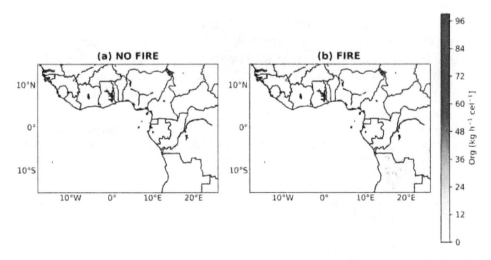

Fig. 2 Biomass burning July fire emissions for the No fire (**a**, 2016) and Fire (**b**, 2016) at the surface

Moderate CO emissions are also observed in the Niger Delta (in southwestern Nigeria), probably as a result of gas flaring activities, as well as in northern Nigeria. The peak of the CO emission is observed in Central and Southern Africa, where the BB is taking place.

Results and Discussion

This section presents the results obtained with regard to BB, by focusing on the impact on meteorology of the study area (section "Impact on Meteorology"), the impact on the atmospheric composition (section "Impact on the Atmospheric Composition"), and the adaptation and mitigation strategies proposed (section "Adaptation and Mitigation Strategies") as being helpful in reducing the impact of BB in the SWA region.

Impact on Meteorology

Our period of investigation is July 2016, which was characterized as the post-onset rainy season in SWA (Knippertz et al. 2017), thus favoring an undisturbed monsoon condition. During this period, long-range transport of BB is advected across the Eastern Atlantic Ocean and the Gulf of Guinea by a jet at around 700 hPa. In order to obtain quantitative results, a domain average over D2 (9°W–4°E; 3°–10.8°N), and the difference and relative change between the Fire and No Fire cases were analyzed. The effects of BB on the meteorological field in general (precipitation, temperature, and cloud droplets number) and on surface shortwave (SW) radiation were examined. Table 1 presents the mean effects over the SWA domain according to the variables for Fire and No Fire.

The results revealed a decrease in SW radiation at the surface and top of atmosphere (TOA) by 6.5 W m^{-2} and 5 W m^{-2}, respectively, a slight decrease in precipitation, and an increase in the Cloud Droplets Number Concentration (CDNC) over SWA. The decrease of surface SW radiation was expected, as aerosols directly reflect or scatter radiation, thus leading to a cooling effect, which in turn reduces the

Table 1 Mean values of meteorological fields for the Fire and No Fire scenarios, the difference (F-NF) and the change in percentage averaged over the nested domain D2 (9°W–4°E; 3°–10.8°N)

Field	Fire (F)	No Fire (NF)	Difference	% change (F-NF) *100/F
Precipitation (mm day^{-1})	4.15	4.26	-0.11 ± 0.008	-2.7
SW down surface (W m^{-2})	83.43	89.95	-6.51 ± 0.0	-7.8
Cloud droplet number (CDNC) concentration (cm^{-3})	94,767	85,542	$10,225 \pm 853$	10.78
TOA net downward SW (W m^{-2})	184.61	190.28	-5 ± 0.0	-3.07

atmospheric stability and, thus, the surface fluxes called semi-direct effects. This result agrees with the findings of Thornhill et al. (2018).

The increase in the CDNC from our model results verifies the Twomey effect (Twomey 1977), which states that an increase in the amount of aerosols leads to a reduction in the cloud droplets size distribution that is associated with an increase in the number concentration.

Our finding corroborates with that of Haslett et al. (2019) which demonstrated that BB aerosols reduce droplet size and increase their number concentration over SWA but less sensitive to precipitation. As a result, more aerosols are competing for water uptake, leading to a reduction in precipitation (known as the Albrecht effect). This result is in agreement with the findings of Thornhill et al. (2018) regarding the impact of BB on precipitation. According to the latter an increase in the number of cloud droplets and a decrease in their size would lead to a decrease in precipitation in the absence of strong convection; however, the decrease here is less compared to their findings.

Impact on the Atmospheric Composition

In addition to investigating the impact of BB on the local climatology of SWA, this study also examined the relative contribution of BB to the atmospheric composition of SWA. The spatial distribution of the surface CO mixing ratio (ppm) is presented in Fig. 3, over both domains, D1 and D2.

The two upper panels (a, b) depict the contribution of BB in CO concentration; it is marked by high values over Central and Southern Africa where BB occurred (Fig. 3b). Low values of traces of CO concentration can be observed over the Atlantic Ocean. The signature of anthropogenic activities is shown by moderate CO values (0.6 ppm) over Lagos and the Niger Delta in southern Nigeria. The simulated CO concentration over D2 in the lower two panels (c, d) clearly illustrate the plumes above the cities (Abidjan, Accra, Cotonou, Lomé, and Lagos); they show moderate values, expect for Lagos, which exhibits the highest CO concentration from both experiments (No Fire and Fire). Kumasi (Ghana), a metropolitan area, also shows a moderate CO concentration as a result of anthropogenic activities. It is worth noting that Kumasi, a city with the third highest population in SWA (3.065 million inhabitants), is one of the most populated area in Ghana (UNO 2018). The CO concentration in SWA cities along the coast is the result of urban pollution. Figure 3d highlights the contribution of BB plumes to the region by showing moderate CO concentrations (0.3 ppm) over the Gulf of Guinea. This is the result of the intrusion of BB air masses into the atmosphere. Indeed, Mari et al. (2008) have shown that BB aerosols are carried aloft and advected westward by a jet at roughly 700 hPa. Layers of BB across SWA, the Gulf of Guinea, and the tropical eastern Atlantic are present during this period of the year (i.e., July 2016) (Chatfield et al. 1998; Mari et al. 2008). The presence of CO from BB in the marine boundary layer mainly arises from subsidence of the aerosols being carried aloft, due to the high pressure area to the west of the African continent (Adebiyi and Zuidema 2016; Flamant et al. 2018). A recent study by Dajuma et al. (2020, in press) demonstrated that

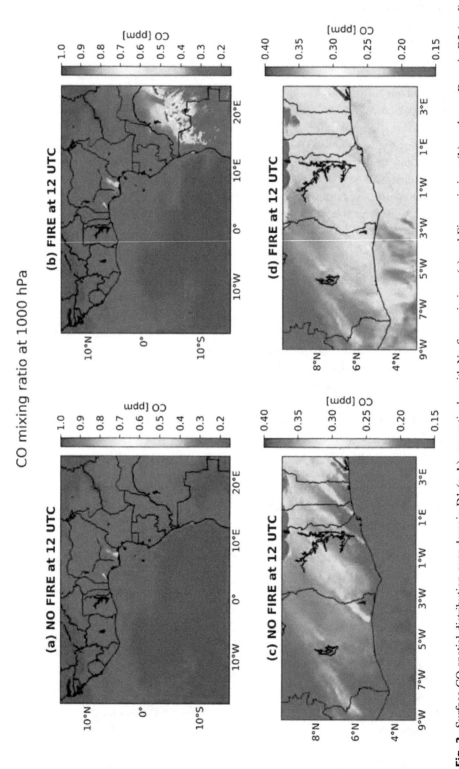

Fig. 3 Surface CO spatial distribution over domain D1 (**a**, **b**), respectively, with No fire emissions (**a**) and Fire emissions (**b**), and over Domain D2 (**c**, **d**), respectively, with No fire emissions (**c**) and Fire emissions (**d**) simulated with COSMO-ART

that convective cumulus clouds located over the Gulf of Guinea play a role in the downward mixing of aerosols located at the mid-level troposphere into the PBL. The southerly monsoon flow in the PBL below 700 hPa carries both BB aerosols and urban gases and aerosols northward (Deroubaix et al. 2019; Knippertz et al. 2017).

The relative contribution of BB to the PM_{25} over urban cities was also examined in our study. Time series of surface concentrations of PM_{25} examined for the period simulated over four SWA cities. The results, as illustrated in Fig. 4, are presented for Abidjan, Accra, Cotonou, and Lagos, for simulations with No Fire, Fire, and the difference between the two scenarios (Fire – No Fire).

The diurnal cycle shows a peak of PM_{25} concentration in the early morning around 6 UTC, which begins to decrease after sunrise. The increase of the PBL height results in a mixing of aerosols, highlighted by a decrease with a minimum value around 15 UTC (Fig. 4a). After sunset, the aerosols start to accumulate again. The impact of BB aerosols appears after only 1 day of simulation (from 6 July 2016). For all four cities, the impact of BB has the same order of magnitude, underscoring the wide extent of long-range BB over SWA. The maximum contribution found was 9.75 μg m^{-3}, 12.45 μg m^{-3}, 10.7 μg m^{-3}, and 8 μg m^{-3}, respectively, in Accra, Abidjan, Cotonou, and Lagos. Menut et al. (2018) quantified the contribution of BB to the PM_{10} concentration to be about 5 μg m^{-3} in two SWA cities (Lagos and Abidjan). Although PM_{25} and PM_{10} are composed of the same type of aerosols, only at different sizes, it is expected that the contribution to PM_{25} will be higher (double) than that of PM_{10}. This confirms that BB aerosols are dominated by the submicron mode, as observed by Haslett et al. (2019).

Our model allowed us to quantify the contributions from each type of aerosol and gases as well. Table 2 thus summarizes the contribution of BB to OC, BC, CO, NO_x, OZONE, and the Aerosol Optical Depth (AOD).

The relation between AOD and aerosol increase is clearly shown. A 36% increase of AOD from BB contribution is simulated by COSMO-ART highlighting the positive relationship between BB aerosols and AOD, in agreement with the findings of (Reddington et al. 2015). Thus, according to our model results, BB significant effect on AOD.

From the analysis of Table 2, it can be seen that BB plumes increased the atmospheric composition of both gases and aerosols concentrations in SWA. The major contribution of aerosols is from OC, representing a 70% increase due to BB, followed by BC, which accounted for 24.5%. These results agreed with the observations made during the DACCIWA campaign, which showed that 80% of the aerosol mass concentration in the monsoon layer (below 1.9 km) in SWA was that originated from Central and Southern African BB (Haslett et al. 2019). Also, Menut et al. (2018), through modeling with the Weather Research and Forecasting model coupled with Chemistry (WRF-Chem), found that the contribution of PM_{10} from BB in Central Africa was mainly composed of primary organic and particulate matter (PM).

Among the gases, the ozone contribution from BB was 16%, followed by CO at 11%, then NO_x at roughly 8%. Real et al. (2010) found that the BB plume over Central Africa was a source of ozone in the mid-level and upper troposphere. Moreover, Mari et al. (2008), tracing BB aerosol from the southern hemisphere

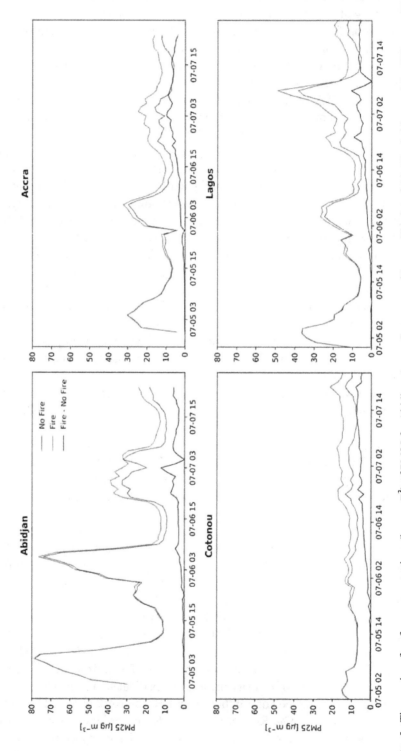

Fig. 4 Time series of surface concentrations (in μg m^{-3}) of PM25 for Abidjan, Accra, Cotonou, and Lagos: Fire in red, No Fire in blue, and the difference (Fire -No Fire) in black

Table 2 Mean values of aerosols mass concentration and gas mixing ratio for the Fire and No Fire scenarios, the difference (F-NF), and the change in percentage averaged over the nested domain D2 (9°W–4°E; 3°–10.8°N)

Concentration	Fire (F)	No Fire (NF)	Difference	% change (F-NF)*100/F
AOD	0.39	0.25	0.14 ± 0.01	36
CO (ppb)	225	199	26 ± 1	11.5
OZONE (ppb)	39	33	6 ± 0.3	16
NO_x (ppb)	26.18	24.1	2.07 ± 0.3	7.98
SOOT ($\mu g\ m^{-3}$)	0.196	0.148	0.05 ± 0.004	24.5
OC ($\mu g\ m^{-3}$)	3.15	0.93	2.21 ± 0.14	70.32

during the summer monsoon, found ozone production to vary, depending on the meteorological conditions associated with the active and break phases of the southern hemisphere jet.

The presence of BB in the PBL increases the level of air of pollution which increases pollution-related diseases (Lelieveld et al. 2015). Intensive measurements during the DACCIWA campaign in both the dry and wet seasons confirmed that carbonaceous aerosols (OC and BC) were major contributions to particle fractions in Abidjan and Cotonou (Adon et al. 2020). The high content of organic aerosol from BB also has adverse effects on population health throughout the region. For instance, a study by Mauderly and Chow (2008) indicated that organic fractions of ambient PM have adverse respiratory and cardiovascular health outcomes.

Adaptation and Mitigation Strategies

The detrimental effects of air pollution are recognized worldwide, particularly in West Africa (Knippertz et al. 2015), as this region appears to suffer most from climate change as a result of anthropogenic activities. Increasing levels of local pollution and pollution from remote aerosol sources, such as BB from Central Africa and mineral dust from the Sahel and the Sahara, are increasing the aerosol burden over the SWA region. Consequently, poor air quality exposes the population to serious health risks, particularly people living in the coastal cities. Since urbanization is increasing worldwide, there is a need to implement effective management plans and policies to improve adaptive capacity and mitigate the effects of anthropogenic induced air pollution in SWA. An air quality monitoring network should be established in SWA in order to regulate air quality throughout the region over the long term, as suggested by the DACCIWA policy brief (Evans et al. 2019). This will also help to spread awareness among the population of areas of high pollution thus reducing the prevalence of respiratory illnesses, such as asthma. Local populations should be encouraged to shift from using wood and charcoal for cooking to instead using electricity or gas in order to reduce local emissions. For instance, Adon et al. (2020), focusing on local sources domestic fires in Abidjan, showed that 75% of the total PM are carbonaceous aerosols.

Regionally, negotiations are necessary between Central Africa and West Africa in order to reduce both regions' BB emissions, as recommended by the DACCIWA policy brief (Evans et al. 2019). It is furthermore important to develop air pollution policies in SWA. Weather forecasting is an essential tool to mitigate air pollution-related effects. There is a need to improve weather forecasting in SWA, not only in the meteorological field but also to account for aerosols, as these interact with the climate.

Another recommendation is to promote the use of renewable energy, such as solar, for use in industries to reduce the consumption of fossils fuels, which are also exacerbating air pollution. With regard to transportation, the traffic fleet in the SWA region is dominated by old cars (Doumbia et al. 2018). Regulations should be implemented to ban the use of old cars and to improve the public transportation systems in SWA, as both are major sources of pollutants.

By means of climate education and the promotion of green lifestyles the local community could be educated, assisted, and encouraged to participate in the mitigation of climate change. And lastly, it is also recommended that increasing observational networks over SWA and improving modeling tools (e.g., cluster) and human capacity would be effective and efficient ways of tackling issues related to climate change.

Conclusions

It has been found in our study that remote pollution from the long-range transport of BB appears to affect the atmospheric aerosol pollution over SWA during summer monsoon. BB plumes originate from Central and Southern Africa as a result of agricultural activities, land management, livestock grazing, and crop production. They are carried westward by a jet toward the tropical eastern Atlantic Ocean, the Gulf of Guinea and SWA in the mid-level troposphere. BB aerosols were shown to increase the concentration of PM over four coastal cities in SWA in the same range of magnitude. Increase in PM_{25} of 12.45 $\mu g\ m^{-3}$, 9,75 $\mu g\ m^{-3}$, 8 $\mu g\ m^{-3}$, and 10.7 $\mu g\ m^{-3}$ was simulated, respectively, for Abidjan, Accra, Lagos, and Cotonou. Observational studies have shown that PM_{25} aerosols measured in the ADF, for instance, have exceeded the limits set by the World Health Organisation (WHO) limit (Djossou et al. 2018).

Relative contributions of BB to gaseous pollutants such as NO_x, CO, and Ozone are, respectively, 7.98%, 11.5%, and 16%. As far as aerosols are concerned, OC is the dominant contributor to the atmospheric composition from BB with a 70% increase, followed by BC with a 24% increase. The AOD change as a result of BB aerosol is estimated at 36%.

Regarding the effect on climate, it was found that SW radiation decreased by 6.5 $W\ m^{-2}$ at the surface and 5 $W\ m^{-2}$ at the TOA. The mean domain average of CDNC increased by 11% from the No Fire to the Fire case scenario, followed by a slight decrease in precipitation. It is worth noting that this experiment was conducted only over a few days. However, the results do emphasize that there is a need to explore further and obtain more robust results by performing long-term simulations. Based on the results obtained with our model, it is obvious that BB aerosols from the southern hemisphere of Africa are interacting with the WAM dynamics; these interactions need to be investigated in future studies.

In addition, there is a need to implement an air pollution monitoring system to tackle health problems related to pollution and increase the adaptive capacity of West Africa to climate change by limiting local pollution. Moreover, there is a need to implement transboundary agreements in policies making, for example, by collaborating with countries in Central Africa to reduce their emissions from vegetation fires, as suggested by the DACCIWA policy brief (Evans et al. 2019).

References

Adebiyi AA, Zuidema P (2016) The role of the southern African easterly jet in modifying the Southeast Atlantic aerosol and cloud environments. Q J R Meteorol Soc. https://doi.org/10.1002/qj.2765

Adon AJ, Liousse C, Doumbia ET, Baeza-Squiban A et al (2020) Physico-chemical characterization of urban aerosols from specific combustion sources in West Africa at Abidjan in Côte d'Ivoire and Cotonou in Benin in the frame of the DACCIWA program. Atmos Chem Phys 20: 5327–5354. https://doi.org/10.5194/acp-20-5327-2020

Athanasopoulou E, Vogel H, Vogel B et al (2013) Modeling the meteorological and chemical effects of secondary organic aerosols during an EUCAARI campaign. Atmos Chem Phys 13:625–645. https://doi.org/10.5194/acp-13-625-2013

Bangert M, Nenes A, Vogel B et al (2012) Saharan dust event impacts on cloud formation and radiation over Western Europe. Atmos Chem Phys 12:4045–4063. https://doi.org/10.5194/acp-12-4045-2012

Barbosa PM, Stroppiana D, Grégoire JM, Pereira JMC (1999) An assessment of vegetation fire in Africa (1981–1991): burned areas, burned biomass, and atmospheric emissions. Global Biogeochem Cycles. https://doi.org/10.1029/1999GB900042

Bowman DMJS, Balch J, Artaxo P et al (2011) The human dimension of fire regimes on earth. J Biogeogr 38:2223–2236. https://doi.org/10.1111/j.1365-2699.2011.02595.x

Chatfield RB, Vastano JA, Li L et al (1998) The Great African Plume from biomass burning: generalizations from a three-dimensional study of TRACE A carbon monoxide. J Geophys Res Atmos 103:28059–28077. https://doi.org/10.1029/97JD03363

Dajuma A, Ogunjobi KO, Vogel H et al (2020) Downward cloud venting of the central African biomass burning plume during the West Africa summer monsoon. Atmos Chem Phys 20:5373–5390. https://doi.org/10.5194/acp-5373-2020

Deetz K, Vogel B (2017) Development of a new gas-flaring emission dataset for southern West Africa. Geosci Model Dev 10:1607–1620. https://doi.org/10.5194/gmd-10-1607-2017

Deetz K, Vogel H, Knippertz P et al (2018) Numerical simulations of aerosol radiative effects and their impact on clouds and atmospheric dynamics over southern West Africa. Atmos Chem Phys 18:9767–9788. https://doi.org/10.5194/acp-18-9767-2018

Deroubaix A, Flamant C, Menut L et al (2018) Interactions of atmospheric gases and aerosols with the monsoon dynamics over the Sudano-Guinean region during AMMA. Atmos Chem Phys 18:445–465. https://doi.org/10.5194/acp-18-445-2018

Deroubaix A, Menut L, Flamant C et al (2019) Diurnal cycle of coastal anthropogenic pollutant transport over southern West Africa during the DACCIWA campaign. Atmos Chem Phys 19:473–497. https://doi.org/10.5194/acp-19-473-2019

Djossou J, Léon JF, Barthélemy AA et al (2018) Mass concentration, optical depth and carbon composition of particulate matter in the major southern West African cities of Cotonou (Benin)

and Abidjan (Côte d'Ivoire). Atmos Chem Phys 18:6275–6291. https://doi.org/10.5194/acp-18-6275-2018

Doumbia M, Toure NE, Silue S et al (2018) Emissions from the road traffic of West African cities: assessment of vehicle fleet and fuel consumption. Energies 11:1–16. https://doi.org/10.3390/en11092300

Edgar (2010) EDGAR – Emission Database for Global Atmospheric Research. Glob Emiss EDGAR v42 (November 2011). https://doi.org/10.2904/EDGARv4.2

Emmons LK, Walters S, Hess PG et al (2010) Description and evaluation of the Model for Ozone and Related chemical Tracers, version 4 (MOZART-4). Geosci Model Dev 3:43–67. https://doi.org/10.5194/gmd-3-43-2010

Evans MJ, Knippertz P, Aristide A, Allan RP (2019) Policy-relevant findings of the DACCIWA. https://doi.org/10.5281/zenodo.1476843

Flamant C, Deroubaix A, Chazette P et al (2018) Aerosol distribution in the northern Gulf of Guinea: local anthropogenic sources, long-range transport, and the role of coastal shallow circulations. Atmos Chem Phys 18:12363–12389. https://doi.org/10.5194/acp-18-12363-2018

Hao WM, Liu MH (1994) Spatial and temporal distribution of tropical biomass burning. Global Biogeochem Cycles. https://doi.org/10.1029/94GB02086

Haslett SL, Taylor JW, Evans M et al (2019) Remote biomass burning dominates southern West African air pollution during the monsoon. Atmos Chem Phys 19:15217–15234. https://doi.org/10.5194/acp-19-15217-2019

Huang J, Adams A, Wang C, Zhang C (2013) Black Carbon and West African Monsoon precipitation: observations and simulations. Ann Geophys 27:4171–4181

Huang X, Ding A, Liu L et al (2016) Effects of aerosol-radiation interaction on precipitation during biomass-burning season in East China. Atmos Chem Phys 16:10063–10082. https://doi.org/10.5194/acp-16-10063-2016

Jiang Y, Lu Z, Liu X et al (2016) Impacts of global open-fire aerosols on direct radiative, cloud and surface-albedo effects simulated with CAM5. Atmos Chem Phys 16:14805–14824. https://doi.org/10.5194/acp-16-14805-2016

Kaiser JW, Heil A, Andreae MO et al (2012) Biomass burning emissions estimated with a global fire assimilation system basedon observed fire radiative power. Biogeosciences 9:527–554. https://doi.org/10.5194/bg-9-527-2012

Knippertz P, Evans MJ, Field PR et al (2015) The possible role of local air pollution in climate change in West Africa. Nat Clim Chang 815–822. https://doi.org/10.1038/nclimate2727

Knippertz P, Fink AH, Deroubaix A et al (2017) A meteorological and chemical overview of the DACCIWA field campaign in West Africa in June-July 2016. Atmos Chem Phys 17:10893–10918. https://doi.org/10.5194/acp-17-10893-2017

Knote C, Brunner D, Vogel H et al (2010) Online-coupled chemistry and aerosols: COSMO-ART modelperformance

Lana A., Bell T. G,. Simó R., et al (2011) An updated climatology of surface dimethlysulfide concentrations and emission fluxes in the global ocean. Global Biogeochem Cycles 25:1–17. https://doi.org/10.1029/2010GB003850

Lelieveld J, Fnais M, Evans JS et al (2015) The contribution of outdoor air pollution sources to premature mortality on a global scale. Nature 525:367–371. https://doi.org/10.1038/nature15371

Marenco A, Medale JC, Prieur S (1990) Study of tropospheric ozone in the tropical belt (Africa, America) from STRATOZ and TROPOZ campaign. Atmos Environ 24:2823–2834. https://doi.org/10.1016/0960-1686(90)90169-N

Mari CH, Cailley G, Corre L et al (2008) Tracing biomass burning plumes from the Southern Hemisphere during the AMMA 2006 wet season experiment. Atmos Chem Phys 8:3951–3961. https://doi.org/10.5194/acp-8-3951-2008

Mauderly JL, Chow JC (2008) Health effects of organic aerosols. Inhal Toxicol 20:257–288. https://doi.org/10.1080/08958370701866008

Mbow C, Nielsen TT, Rasmussen KC (2000) Savanna fires in East-Central Senegal: distribution patterns, resource management and perceptions. Hum Ecol 28(4):561–583

Menut L, Flamant C, Turquety S et al (2018) Impact of biomass burning on pollutant surface concentrations in megacities of the Gulf of Guinea. Atmos Chem Phys 18:2687–2707. https://doi.org/10.5194/acp-18-2687-2018

N'Datchoh ET, Konaré A, Diedhiou A et al (2015) Effects of climate variability on savannah fire regimes in West Africa. Earth Syst Dynam 6:161–174. https://doi.org/10.5194/esd-6-161-2015

Pani SK, Wang SH, Lin NH et al (2016) Radiative effect of springtime biomass-burning aerosols over northern Indochina during 7-SEAS/BASELInE 2013 campaign. Aerosol Air Qual Res 16:2802–2817. https://doi.org/10.4209/aaqr.2016.03.0130

Real E, Orlandi E, Law KS et al (2010) Cross-hemispheric transport of central African biomass burning pollutants: implications for downwind ozone production. Atmos Chem Phys 10:3027–3046. https://doi.org/10.5194/acp-10-3027-2010

Reddington CL, Butt EW, Ridley DA et al (2015) Air quality and human health improvements from reductions in deforestation-related fire in Brazil. Nat Geosci 8:768–771. https://doi.org/10.1038/ngeo2535

Stowe LL, Jacobowitz H, Ohring G et al (2002) The advanced very high resolution radiometer (AVHRR) pathfinder atmosphere (PATMOS) climate dataset: initial analyses and evaluations. J Clim 15:1243–1260. https://doi.org/10.1175/1520-0442(2002)015<1243:TAVHRR>2.0.CO;2

Swap RJ, Annegarn HJ, Suttles JT et al (2002) The southern African regional science initiative (SAFARI 2000): overview of the dry season field campaign. S Afr J Sci 98:125–130

Thornhill GD, Ryder CL, Highwood EJ et al (2018) The effect of south American biomass burning aerosol emissions on the regional climate. Atmos Chem Phys 18:5321–5342. https://doi.org/10.5194/acp-18-5321-2018

Twomey S (1977) The influence of pollution on the shortwave albedo of clouds. J Atmos Sci 34:1149–1152. https://doi.org/10.1175/1520-0469(1977)034<1149:TIOPOT>2.0.CO;2

UNO (2018) The world's cities in 2018. https://doi.org/10.18356/c93f4dc6-en

Vogel B, Vogel H, Bäumer D et al (2009) The comprehensive model system COSMO-ART – radiative impact of aerosol on the state of the atmosphere on the regional scale. Atmos Chem Phys 9:8661–8680. https://doi.org/10.5194/acp-9-8661-2009

Walter C, Freitas SR, Kottmeier C et al (2016) The importance of plume rise on the concentrations and atmospheric impacts of biomass burning aerosol. Atmos Chem Phys 16:9201–9219. https://doi.org/10.5194/acp-16-9201-2016

Zängl G, Reinert D, Rípodas P, Baldauf M (2015) The ICON (ICOsahedral non-hydrostatic) modelling framework of DWD and MPI-M: description of the non-hydrostatic dynamical core. Q J R Meteorol Soc 141:563–579. https://doi.org/10.1002/qj.2378

7

Multifunctional Landscape Transformation of Urban Idle Spaces for Climate Resilience in Sub-Saharan Africa

David O. Yawson, Michael O. Adu, Paul A. Asare and Frederick A. Armah

Contents

Introduction ....
    The Challenges of Urbanization, Climate Change, and Resilience in Africa
    Evolution of Urban Landscapes and Idle Spaces in Africa
    Multifunctional Edible Landscape Approach to Resilience
    Structure of the Chapter
Multifunctional Edible Urban Landscape Transformation in Practice
    Implementation of the Pilot Project
    Project Achievement
    Lessons and Insights for Scaling Up
Summary and Conclusions
References

Abstract

Poor physical and land use planning underpin the chaotic evolution and expansion in cities and towns in sub-Saharan Africa. This situation amplifies urban vulnerability to climate change. Worse, urban landscapes are rarely considered part of the discourse on urban development in sub-Saharan Africa, let alone in climate change adaptation. Yet, landscapes are known to play crucial roles in social, economic, and cultural resilience in cities and towns. Hence, designing basic forms of appealing and functional urban landscapes that support multiple

D. O. Yawson (✉)
Centre for Resource Management and Environmental Studies, The University of the West Indies, Bridgetown, Barbados
e-mail: david.yawson@cavehill.uwi.edu

M. O. Adu · P. A. Asare
Department of Crop Science, University of Cape Coast, Cape Coast, Ghana
e-mail: michael.adu@ucc.edu.gh; pasare2@ucc.edu.gh

F. A. Armah
Department of Environmental Science, University of Cape Coast, Cape Coast, Ghana
e-mail: farmah@ucc.edu.gh

ecosystem services is essential to the drive towards resilience, which relates to the ability to maintain or improve the supply of life support services and products (such as food and water) in the face of disturbance. In this chapter, the idea of transforming idle urban spaces into multifunctional edible urban landscapes is introduced and explored as instrumental for cost-effective adaptation and resilience to climate change in cities and towns in sub-Saharan Africa. *Multifunctional edible urban landscape* is defined here as a managed landscape that integrates food production and ornamental design, in harmonious coexistence with other urban structures to promote or provide targeted, multiple services. These services include food security, scenic beauty, green spaces for active living and learning, jobs and livelihoods support, environmental protection, climate adaptation, and overall urban resilience. This approach constitutes a triple-win multifunctional land use system that is beneficial to landowners, city managers, and the general community. This chapter explores the benefits, challenges, and prospects for practically transforming urban idle spaces into multifunctional edible urban landscapes using an example project from Ghana. The chapter shows that multifunctional edible urban landscape transformation for resilience is practically feasible, and sheds light on the possibility of the food production component paying for landscaping and landscape management. It concludes with thoughts on actions required across sectors and multiple scales, including mobilizing stakeholders, laws, policies, and incentives, to actualize multifunctional edible urban landscapes as key transformational components of resilience in sub-Saharan Africa.

Keywords

Urban landscape · Multifunctional land use · Resilience · Green spaces · Food security · Ecosystem services · Climate change adaptation

Introduction

The Challenges of Urbanization, Climate Change, and Resilience in Africa

Urbanization has its promise and value in terms of concentrating resources and capital to improve physical development and human well-being; but it has its perils too as it increases the demand for basic necessities of life (such as food, water, shelter, employment, and recreational opportunities) and the complexity of managing the resources and the socioecological processes that underpin ecosystem services and quality of life. Urbanization creates and alters landscape attributes and functions, which in turn affect the environment and quality of life in urban areas. Compared to the rate of urbanization in other regions, Africa is said to have achieved an urbanization miracle (Lwasa 2014). Africa has experienced an unprecedented rate of urbanization in the last two decades, resulting in the emergence of several megacities

(Lwasa 2014; Güneralp et al. 2017). From the year 2010, African urban population has increased at an average rate of 3.2% per year, compared to global average of about 1.2% (African Union 2020; UN DESA 2012). Africa currently has 43% urban population and, together with Asia, will account for about 90% of the projected increase in the world's urban population by 2050 (UN DESA 2019). By mid-century, more than half of Africa's population is expected to live in urban areas. Sub-Saharan Africa will host most of the largest cities that will arise from the projected population growth in Africa, most of them in coastal areas (Di Ruocco et al. 2015). These underscore the challenge and importance of sustainably managing the quality and resilience of urban environment and life by maintaining and enhancing the integrity and functionality of the systems that underpin resilience, particularly food systems (Russo et al. 2017; Haberman et al. 2014).

Already, cities and urban communities in Africa face formidable challenges, including high levels of food insecurity, unemployment, poverty, and environmental degradation (African Union 2020; Lwasa 2014; Güneralp et al. 2017). Provision of livable urban spaces, including green and recreational spaces, has proved particularly challenging and elusive. Increase in food production in Africa has substantially lagged behind the rate of population growth, resulting in protracted food insecurity (African Union 2020). The number of hungry people in Africa is estimated at 256 million and the continent might not meet key global targets on malnutrition or eliminate hunger by 2030 (FAO, ECA and AUC 2020). Projections suggest that an increase of about 112% in food production (over 2015 baseline) will be required to meet the food demand of sub-Saharan Africa alone in 2050 (African Union 2020). Urbanization will not only increase the quantity of food demanded, but also alter the composition and patterns of food consumption, with likely increase in demand for fresh fruits and vegetables. At the same time, urbanization will increase competition with agriculture for land and water which underpin food production.

Climate change presents an additional developmental burden and complicates existing challenges for all regions. However, Africa is considered a highly vulnerable continent to the adverse impacts of climate change (IPCC 2007; Chapman et al. 2017), and the continent is already experiencing the impacts of climate change, through highly frequent and severe episodes of droughts and floods (African Union 2020; Batchelor and Schnetzer 2018; Van Rooyen et al. 2017; Lwasa 2014; Armah et al. 2011). This vulnerability stems partly from Africa's poor physical planning and environmental degradation, and overreliance on agriculture, which is hypersensitive to climate change (IPCC 2007). These challenges raise the imperative for exploring several response options, including the planning, evolution, development, and uses of urban landscapes for reducing vulnerability and enhancing resilience (Russo et al. 2017).

Evolution of Urban Landscapes and Idle Spaces in Africa

Urbanization can be instrumental in amplifying vulnerability or enhancing resilience. Unfortunately, urbanization in Africa is rapid, unplanned, unregulated, and chaotic (Lwasa 2014; Güneralp et al. 2017), a situation that will compound vulnerability. Especially in sub-Saharan Africa, physical development and land use are poorly planned (if at all). Weak land administration and poor municipal services add to the challenge of discontinuous expansion and poor-quality urban landscapes and environments (Mensah 2014). Urban landscape planning and management are rarely considered important in the discourse on sustainable and resilient development (Yawson 2020). Urban landscapes have, therefore, evolved organically and chaotically with urbanization, and are poorly managed. This poor planning and management of urban landscapes has resulted in the degradation of the biophysical and sociocultural significance of urban communities and diminished the overall resilience to environmental shocks. If this trend remains unaddressed, it can substantially amplify urban challenges and vulnerability to climate change (Di Ruocco et al. 2015). One visible result is the considerable proportion of land surfaces poorly covered by either natural or man-made covers (Mensah 2014). Consequently, very dusty, bushy, and unsightly land surfaces are visible in both the urban core and the periphery, resulting in high exposure to dust during the dry season and mud or sediment transport during the wet season, and other unhealthy elements (Yawson 2020). These have consequences for resilience and sustainability in both the short and long term. For example, long-term exposure to $PM_{2.5}$ (dominated by dust) can contribute to ischemic heart disease, cerebrovascular disease, lung cancer, and respiratory diseases (Health Effects Institute 2019). The number of deaths associated with $PM_{2.5}$ in Ghana and Nigeria in 2017 were 5,190 and 49,100, respectively (Health Effects Institute 2019).

Land ownership in Ghana can be used to illustrate the evolution of urban idle spaces in urban landscapes in sub-Saharan Africa. In Ghana, just as in almost all sub-Saharan African countries, urban expansion occurs largely through private action in an informal context. There are two broad types of land ownership in Ghana: customary and state ownership, accounting for 78% and 20%, respectively of total land (Larbi 2008; Yawson and Armah 2018). The remaining 2% are vested lands (split ownership between the state and customary authorities). Customary lands are owned by traditional authorities (stools and skins), clans and families, with chiefs and clan/family heads as the custodians and supported by principal elders of the given community or family (Larbi et al. 2004). The government of Ghana can apply the power of eminent domain to acquire land for national development or security purposes. It is believed that this system of land ownership constrains effective land administration, land use planning and management, and is a major driver of informal land transactions and development. It also partly accounts for the numerous intractable land disputes and conflicts from parcel, through community to tribal scales. Although the Town and Country Planning Department is responsible for physical planning, chiefs or owners of large tracts of land are expected to make and present a parcelled plan of their lands to the Lands Commission for the purposes of registration and titling. However, this is not mandatory or enforced. This means parcelling and layout of residential areas largely evince from landowners, independent of infrastructural and municipal service planning, delivery, and management. This results in incoherent physical development, with scattered pattern of settlements in developing areas. Due to weak property markets, insecurity of tenure of residential or commercial properties, and weak financial support system for rapid development of individual properties, there is inordinate pressure on land for low-density housing and other commercial activities in urban areas. As a result, persons buy parcels of land and slowly develop these, or simply leave the parcels of land idle until they are financially capable of developing the land. Some also buy land out of speculation, with the view to selling the land later at higher prices. Undeveloped parcels of land do not attract property taxes and there are no mechanisms to ensure management regimes fit for the landscape or the environment. As a result, and partly due to informal planning of space, there are several open spaces and undeveloped or partly developed but idle parcels of land that are poorly managed (if at all) in both the urban core and periphery. These, together with poor infrastructural planning, management, and municipal services, diminish the natural, cultural, and scenic beauty of urban landscapes and escalate vulnerability.

These idle parcels do not only detract from the urban beauty and landscape services, but are also sites for a range of activities, both positive and negative. In some cases, these parcels are used by itinerant farmers for food production, occupied by petty traders or informal settlers (squatters), or simply left bushy or bare and exposed to the erosive forces of water and wind (dust generation and transport). In most cases, these parcels are used for nefarious activities such as littering, defecation, and criminal activities, or habitat for vermin. However, these spaces can be sites for urban greening and human production to transform the landscape to support multiple environmental and developmental goals such as jobs or income generation, food security, environmental protection, and scenic beauty, and thereby contribute to overall urban resilience and well-being.

Multifunctional Edible Landscape Approach to Resilience

Resilient development and adaptation planning have become more urgent for social groups, ecological systems, and geographic regions for which there is high certainty of vulnerability (Stern and Treasury 2007). Resilience has been conceptualized as the capacity of a dynamic system to maintain its structural and functional integrity, or change for the better, in the face of disturbances that threaten the viability, functioning, or development of that system (Masten 2014). Urban resilience, from a socio-ecological perspective, removes constancy and introduces flexible, adaptive, and multilevel approaches for responding to persistent or short-term threats and stresses across the entire urban space. The scale of challenges outlined in section "The Challenges of Urbanization, Climate Change and Resilience in Africa" constitute a major source of disturbance that threatens the viability, functioning, and development of urban landscapes and communities, deserving resilient, multilevel responses. Globally, landscape-based approaches are being promoted, in both policy and academic circles, as an integral component of the responses to urban sustainability challenges (Säumel et al. 2019; van Noordwijk et al. 2014; Russo et al. 2017; Panagopoulos et al. 2016; Matsuoka and Kaplan 2008).

Landscapes can have different meanings for different persons or professionals (Fischer et al. 2016; Scott et al. 2009). However, in the urban context, landscapes can be considered as the physical environment shaped by human-nature interaction. The "nature" component comprises the biophysical cover of the land surface that underpin ecological productivity, while the "human" component comprises man-made surfaces and structures such as houses, roads, and other infrastructure (Grêt-Regamey et al. 2015; Panagopoulos et al. 2016; Russo et al. 2017), together with human action on nature. These components are intertwined and provide reciprocal services and or feedback effects on each other. Landscapes, therefore, provide first impressions of the complex processes and interactions between social and ecological components that underpin the provision of ecosystem services, which, in turn, underpin human well-being and resilience (Yawson 2020). These ecosystem services are classified as provisioning (e.g., food, herbs, and clean water or air), supporting (e.g., nutrient, energy, or material cycling), regulatory (e.g., flood, erosion, and climate control), and cultural (e.g., heritage, spirituality, visual aesthetics, and recreational opportunities) (Grêt-Regamey et al. 2015). A landscape reflects the status of cultural sophistication, development, and well-being, as well as the vulnerability of its inhabitants or users to a given external or environmental shock (Yawson 2020). Landscapes that are subject to unsustainable management practices can be very vulnerable to shocks and, in turn, deepen the vulnerability of its inhabitants or users to environmental shocks.

Multifunctional edible urban landscapes are being promoted as instrumental for sustainably intensifying the production of ecosystem services and products, improving livelihoods, and optimizing the use of limited land resources as opposed to straight intensification of farming and commercial forestry (Säumel et al. 2019; Panagopoulos et al. 2016; Fischer et al. 2016; Bustamante et al. 2014; van Noordwijk et al. 2014; Santika et al. 2015). Multifunctional edible urban landscape refers to a managed landscape that integrates food production and ornamental landcover, in harmonious coexistence with other urban structures to promote or provide targeted, multiple services such as food, erosion and flood control, scenic beauty, and recreation. A multifunctional edible urban landscape approach, thus, helps generate ecosystem goods and services to meet the basic but multiple demands of urban communities while limiting expansive use of land. It helps reconcile competing interests and goals that put urban land under constant and intense pressure. Multifunctional edible urban landscapes will arise from and be enhanced by land use planning approaches and innovative transformations that harmonize the functions of landscape components in order to derive multiple benefits across varying spatial and temporal scales. Multifunctional edible urban landscapes have considerable promise to contribute to urban resilience and long-term sustainability or adaptive capacity if the structure and functions of components are well aligned, interconnected, and harmonized to generate specific products and services (Säumel et al. 2019; Panagopoulos et al. 2016; van Noordwijk et al. 2014). This implies urban planning should be deliberate in shaping multifunctional landscape performance in order to derive optimal outcomes from every inch of ground. This can be enhanced or constrained by land use intensity and the scale of opportunity for landscape innovations in existing urban zones. It could be argued that several aspects of landscapes are multifunctional in nature. However, developing countries require special attention for multifunctional landscape planning to impose desirable landscape structure with increased capacity to contribute to addressing the multiple challenges in urban communities in the short term, while enhancing resilience and adaptive capacity in the long term.

Increasingly, this realization has given rise to the calls for expanded urban green infrastructure as a nature-based solution or ecosystem-based approach to resilience and climate change adaptation (Russo et al. 2017). Urban greenspace or cover can contribute to food security, urban aesthetics, thermal regulation, air quality (by suppressing dust and removing pollutants), water quality, and mitigate against flood and erosion (Wolch et al. 2014; Eschobedo et al. 2011; Thompson 2011; Nowak et al. 2006). A review by Laille et al. (2014) showed a strong evidence for contribution to physical and psychological health, biodiversity, thermal regulation, and urban attractiveness. In fact, poor access to urban greenspace could be associated with increased mortality (Coutts et al. 2010) and other adverse health outcomes (e.g., Villeneuve et al. 2012; Thompson 2011; Barton and Pretty 2010). This emerging evidence supports the view on the value and utility of urban green cover to strengthen resilience and adaptation to climate change (Dai 2011). Accordingly, research interests in green infrastructure, ecosystem services, or nature-based solutions have increased substantially as options for strengthening urban sustainability and resilient response to climate change (Haase et al. 2014). While a large body of literature has emerged from this research interest, these studies are largely conducted in isolation without considering the need for integrated systems that include food production and general landscape management in urban contexts (Russo et al. 2017). Critically, in developing countries, the role of human production activities as part of green cover in urban landscapes to support multiple goals, including support for livelihoods, food security, environmental protection and aesthetics, and provision of recreational and learning opportunities, needs to be emphasized (Yawson et al. 2019).

Elsewhere, there have been calls for embedding multifunctionality into agriculture and other land use sectors in rural landscapes (Mander et al. 2007; Nair and Garrity 2012). Even though interest in urban agriculture has been increasing substantially (Csortan et al. 2020; FAO 2012; Urban Agriculture 2009), this has often been studied or articulated in isolation from general landscape management goals. In Africa, the need for deliberately incorporating human production activities in urban landscapes to serve multifunctional purposes is urgent and critical due to increasing urban poverty, food insecurity, and poor-quality landscapes (FAO 2012). Due to poor urban planning, landscapes evolve chaotically with urban development and are hardly managed (Lwasa 2014; Mensah 2014). Poor financial resources and land administration combine to raise the challenge of creating and managing curated landscapes to preserve the scenic beauty and improve the social, economic, environmental, and cultural significance of urban zones. A way out is mobilizing policies, regulations, incentives, and stakeholders to adopt landscape management approaches that permit a sensible balance between economic activities and profitability on one hand, and ecological functionality and productivity on the other hand, so that the latter is maintained by the former and, together, support overall urban well-being and development goals. In other words, producers become landscape entrepreneurs and managers whose obligation include the maintenance and management of ecological productivity and functionality, profitability, and scenic beauty. In this case, the scenic beauty, which is a desirable public good or noncommodity output, for example, is paid for by the economic production which is private activity (Mander et al. 2007). This twin system can be applied to multifunctional landscapes where food production supports urban food security, jobs, and poverty reduction, while general landscape management, as part of the production obligations, provides ecosystem services such as scenic beauty, environmental protection, and recreational opportunities.

In Africa, innovations in existing urban spaces are required to improve livability and overall resilience and human well-being, through increased capacity for intensive delivery of multiple ecosystem services. Because land is finite and the value of urban land parcels increases rapidly, preserving spaces for a single purpose is extremely difficult (Säumel et al. 2019; Güneralp et al. 2017; Mander et al. 2007). This difficulty is heightened by poor land administration, speculation, informal land transactions, and urban expansion. Additionally, the constraint of funding makes it difficult for African governments to invest in the development and maintenance of curated landscapes to improve the scenic beauty of the urban space and contribute to resilience and well-being. As a result, improving landscape contribution to resilience would require approaches that balance or optimize ecological productivity through human production, economic profitability, and scenic beauty or delivery of other ecosystem services which are public goods. To this end, it is important to adopt innovative approaches to impose specific multifunctional landscape character in existing urban spaces, and multifunctional land use planning for the future. Multifunctional landscape approaches that include human production are promising options for enhancing resilience in poorly planned urban zones in Africa (Säumel et al. 2019). Particularly, food production integrated in properly designed and managed landscapes to support multiple development goals, including food security and jobs, would be crucial. This is particularly important since land is finite, urban land has high value and is constantly under intense pressure from competing interests and goals. Multifunctional edible landscape approach provides opportunity for balancing competing interests and goals, both ecological and socioeconomic, in the African urban space. However, a multistakeholder mobilization of policies, resources, and regulatory and financial instruments will be required to ensure inclusivity, broad-based acceptance, and long-term sustainability (Yawson et al. 2019). The purpose of this chapter is to illustrate the workability and utility of this innovative approach to urban landscape transformation, and the requirements for scaling out, for resilience in Africa using a case study from Ghana.

Structure of the Chapter

Section "Introduction" above has provided background and contextual information about urban challenges and evolution of urban landscapes in Africa, as well as the utility of mutifunctional edible landscapes for enhancing resilience. The rest of the chapter is organized as following. Section "Multifunctional Edible Urban Landscape Transformation in Practice" presents a case study of practical implementation of multifunctional edible landscape in existing urban areas in Ghana and the outcomes. Section "Lessons and Insights for Scaling Up" builds on the case study outcomes to provide lessons and insights on practical expansion of multifunctional edible urban landscapes in Africa. Finally, the chapter presents some conclusions and recommendations in section "Summary and Conclusions".

Multifunctional Edible Urban Landscape Transformation in Practice

Promoting innovative landscape transformation can play a catalytic role in addressing urban challenges and enhancing resilience to climate change in Africa. Proceeding from the context and belief presented in the prior section, a practical example of an edible urban landscape transformation activity Ghana is presented in this section. This example is aimed at demonstrating the utility and practical approach to transforming idle spaces into multifunctional edible landscapes in existing urban areas. It also highlights key challenges and levers of change for wider implementation or adoption of this approach to support multiple development goals and enhance resilience. A question of interest inherent in this example was whether the food production component, from a multifunctional perspective, could incentivize and pay for maintenance of scenic beauty and management of landscapes in existing urban areas.

The practical example referred to is derived from a pilot project by the Agriculture for Food Security 2030 (AgriFoSe2030, theme 2) sponsored by the Swedish International Development Agency (SIDA) through the University of Gothenburg. The project, firstly, tested the idea of using vacant, idle parcels of land in the urban area to produce food, enhance the scenic beauty of the surrounding landscape with ornamental plants, while providing jobs for young people and women. This concept was referred to as multifunctional edible urban landscape (a managed landscape that integrates food production, ornamental aesthetics, and other urban structures in a harmonious coexistence to deliver targeted, multiple ecosystem services – Yawson et al. 2019). The edible urban landscape pilot project was conceived as a potential path to climate adaptation and resilience in urban centers in Ghana. The pilot project took place in the year 2018 in the City of Cape Coast in the Central Region of Ghana. Several idle parcels suitable for the project were identified in Cape Coast. Owners of the identified parcels were contacted to negotiate permission to use their lands for the pilot project. Other considerations for the final selection of sites included trust of landowner, safety of the project activities and assets (for e.g. safety from praedial larceny), access to water, and ease of monitoring by the researchers. Based on these, two sites were eventually secured at Akotokyer (1°17′36.28″W, 5° 8′8.61″N) and Kwaprow (1°18′7.02″W, 5° 7′26.98″N), two communities that border the University of Cape Coast.

The city of Cape Coast is the capital of the Cape Coast Metropolis (CCM) and the Central Region of Ghana. The CCM covers an area of 122 km^2 and is very urbanized as only 23% of its 169,894 inhabitants live in rural areas (Ghana Statistical Service 2013). The study communities are among several that border the University of Cape Coast and, together, make up the largest spatially distinct continuum of communities outside of the core of the city of Cape Coast which is densely built up and has limited space for new development. Due to the proximity to the University of Cape Coast, Kwaprow and Akotokyer are among the fastest-growing urban communities in Cape Coast as demand for facilities and services for students, staff, and the associated itinerant workers of the University of Cape Coast keeps growing. However, the communities lack planned physical development and so the physical expansion is haphazard and chaotic, with spatially scattered property development and several idle, unmanaged spaces contributing to undesirable landscapes. A section of urban landscape in Kwaprow is shown in Fig. 1. The communities have poor infrastructure and municipal services are vulnerable to floods and are highly exposed to dust pollution during the dry season. This situation presents an opportunity to generate evidence for the value and feasibility of landscape transformation to support multiple development goals and enhance resilience and well-being.

Fig. 1 A Google Earth screenshot of landscape view of part of Kwaprow community (one of the pilot project communities) in Cape Coast

Implementation of the Pilot Project

The project goals and activities were explained to relevant stakeholders. The sites for the pilot project were bushy and unkempt (Fig. 2). The site at Akotokyer had sparsely distributed apartment buildings while the site at Kwaprow had neighboring low-density houses and amenities such as a school and a clinic. The sites were cleared, ploughed, and harrowed (Fig. 3).

A greenhouse, measuring 9×15 m, was installed at each site (Fig. 4). Solar-powered fans were used for ventilation and to control humidity in the greenhouses. Seedlings of the tomato genotype Eva Purple Ball were grown in pots in the

Fig. 2 Project site at (**a**) Akotokyer and (**b**) Kwaprow

Fig. 3 Site at Kwaprow (**a**) and Akotokyer (**b**) ploughed and prepared for greenhouse installation

Fig. 4 Greenhouses installed at sites with grasses and border plants grown around the greenhouses

Fig. 5 Young tomato plants growing in pots in the greenhouses

greenhouses and recommended agronomic practices on integrated production and protection (IPP) were used (Figs. 5 and 6). The area surrounding the greenhouses were planted with grasses and border plants (fruit trees for shade were planned but land lease terms did not permit this immediately) and maintained to improve the scenic beauty of the site, control erosion and dust, while providing recreational and educational opportunities for neighboring families (Fig. 4). Eight young people, four females, and four males aged between 18 and 35 years were employed to work in the greenhouses and to maintain the surrounding landscapes.

Fig. 6 Matured tomato plants in the greenhouses

Fig. 7 Degraded, idle backyard space (left) turned into a productive site for eggplant (right)

In addition, the project encouraged individuals to transform small spaces in backyards or urban areas into green, productive areas. For example, one woman turned a degraded, idle backyard space into a productive site for eggplants (Fig. 7).

When the tomatoes were ripe and partly harvested, a stakeholder engagement or dissemination event was held at the sites, involving the communities and other stakeholders, to showcase and discuss the relevance, challenges, and opportunities for scaling up the project (Fig. 8).

Fig. 8 Segments of stakeholder dissemination event

Project Achievement

Multifunctional Edible Urban Landscape

The association between urban greenspaces and environmental quality and human well-being is well established and the need for planning to incorporate greener spaces in urban areas is being actively promoted (Panagopoulos et al. 2016; Laille et al. 2014; Mensah 2014; Bratman et al. 2012; Matsuoka and Kaplan 2008). However, in Ghana, and largely across sub-Saharan Africa, outdoor landscaping is a private matter and overall management of public spaces and the general fabric or structure of the urban landscape does not feature in discourse on sustainable development or resilience. There is evidence that greenspaces in urban Africa is declining at an alarming rate due to neglect and poor land use planning and controls (Mensah 2014). Poor urban landscapes can diminish resilience as they can contribute to flooding and sediment transport, thermal stress, air pollution, garbage accumulation, and transmission of diseases. Landscapes expose most urban residents to environmental conditions that undermine health, well-being, natural resource sustainability, and overall resilience (Säumel et al. 2019). It has been reported that the informal, chaotic, and discontinuous pattern of urbanization in Africa, together with poor social services, is a mark of vulnerability to extreme events (Di Ruocco et al. 2015). Dust, for example, is a major component of particulate matter in the air. In most urban zones and rural communities in Africa, people live literally in dust arising not only from industrial or human activities but mainly from bare surfaces (including roads, see Fig. 1) due to poor physical, infrastructural, and landscape planning and management. While the adverse health outcomes of exposure to dust are known and have been of considerable interest in some jurisdictions (Khan and Strand 2018), there is poor information on the contribution of dust to air quality and human health in West Africa, for example, where dust pollution and morbidity rates are higher (De Longueville et al. 2010) and over 20% of infant mortality is due to respiratory infections (Bryce et al. 2005; Morris et al. 2003). In the recent state of global air report, it was shown that annual death from air pollution (principally from $PM_{2.5}$) is on the rise in sub-Saharan Africa, reaching 5,190 and 49,100 for Ghana and Nigeria, respectively, in 2017, while others suffer various morbidities due to long-term exposure (Health Effects Institute 2019). Malaria is a major cause of morbidity in the case study communities. Africa remains the largest global hotspot of death from malaria, accounting for about 90% of all deaths from malaria in 2017 (WHO 2018). In addition, heat stress events would likely increase in intensity, frequency, duration, and spatial spread (Chapman et al. 2017) and humid tropical countries could have high sensitivity and exposure to these events. Enhancing resilience in urban communities implies taking verifiable steps to minimize the exposure of populations to these environmental stressors and hazards. Innovative landscape management approaches can help increase infiltration, reduce floods and sediment transport, and exposure to dust and vector-borne diseases such as malaria. These conditions are expected to be amplified by climate change; and taking remedial measures now through cost-effective landscape transformation or management approaches, as demonstrated in this example project, is a reasonable investment in achieving multiple development goals and enhancing resilience and long-term adaptive capacity.

Through their natural, cultural, and scenic beauty, urban landscapes provide a range of services (or disservices) that contribute to social, economic, and cultural resilience and overall quality of life. Landscapes embody and reflect the state of well-being and vulnerability of inhabitants to shocks. Hence, designing basic forms of appealing and functional urban landscapes that support multiple ecosystem services is essential to the drive towards resilience, which relates to the ability to maintain or improve the supply of life support services and products (such as food and water) in the face of disturbance. Just like many cities and towns in Ghana, the evolution of Cape Coast can be described as chaotic, contingent, discontinuous, and informal. Poor planning has combined with pressure on land for housing to create discontinuous physical expansion of the city, resulting in several vacant, idle parcels of land in the city. Several areas in both the core and periphery of the city are degraded, with visible signs of erosion, red-earth (iron-rich) dusty surfaces. Other places are simply covered by unmanaged bush, which becomes breeding grounds for mosquitoes and other vermin. Because of poor drainage systems, the project com-munities easily succumb to floods and sediment transport (Tham-Agyekum et al. 2019), and have conducive surfaces for breeding of mosquitoes. In the example project presented, the sites acquired for the pilot project were idle, bushy, and unmanaged (Fig. 2), detracting from the scenic beauty and ecological utility of the area, as well as posing physical and health hazards to residents and those who passed through the area, especially women, at night. Residents and those who pass through the area harbored fear about the high probability of attacks from criminals and/or vermin. There were signs of nefarious use of the sites as litter, human faecal matter, and other vermin were observed during land preparation. This pilot project opened up the area, improved the ecological functions, safety, scenic beauty, and social utility of the area (Figs. 3 and 4). The landscaping around the greenhouses were generally part of the integrated preventive environmental strategy of beautifying and protecting the landscape, reducing environmental degradation by ensuring soil cover and increasing the delivery of ecosystem services, both directly and indirectly. The idea of the project was to maintain a beautiful landscape or urban aesthetics while producing food and supporting incomes for the landscape managers. In this way, food production and basic landscaping are integrated, in a multifunctional edible urban landscape perspective, to deliver multiple ecosystem services. The project sites became attractions to the residents, the communities, and others who simply passed through the area. The grassed surroundings were used for recreational purposes by families and for curiosity learning (especially by children). The project sites became the largest curated, managed outdoor area considered by residents as beautiful and safe space for recreation by children in the communities. Generally, residents and those who passed by were happy about the project as they realized that it is possible to produce so much food in a small area and, at the same time, maintain a beautiful landscape or communal outdoor space in hitherto idle, unmanaged urban areas. The experience from the pilot project reported in this chapter suggests that persons have innate desire for beautiful urban landscapes but are denied of these due to poverty of municipal services. It also suggests the existence of a desire for recreational opportunities in safe, well-kept urban landscapes due to the use of the site for recreational purposes. Some nearby residents were happy because the improved landscape also meant improved security and safety in the area. In all, the example project demonstrated the idea and utility of multifunctional edible urban landscape and the feasibility of achieving this through transformation of idle spaces in existing urban areas.

Urban Food Security and Livelihoods

Food insecurity, unemployment, and poverty are major developmental challenges in urban Africa. Urban poverty is increasing in Africa and so is food insecurity (African Union 2020) as urban areas rely on rural food supplies. With poor supply chains, fresh food deteriorates rapidly both in transit and in urban markets and food safety also becomes a concern. This constrains access to fresh food as availability and prices become affected. Tomato, for example, is a climacteric crop which is an essential ingredient in almost every Ghanaian dish. Tomato is produced under rainfed conditions and long-distance transport and storage under poor conditions result in high losses. Prices of tomato in the dry season can be as high as three to four times the prices in the rainy or main harvesting season. This raises the need for producing food close to urban markets or point of consumption. Although urban agriculture is practiced in Ghana, there are concerns about the safety of the food produced and its environmental impacts due to the quality of water, other inputs, and management practices used (FAO 2012; Redwood 2009). Because urban agriculture has yet to gain a desirable level of quality and policy attention and thereby become mainstreamed in the urban landscape architecture, urban agriculture is practiced informally and in isolation from landscape management goals (FAO 2012), let alone from a multifunctional perspective. Just as with food, even where municipal services are available, poverty constrains access to these, resulting in increased vulnerability (Armah et al. 2018). Overall, there is a need to boost policy and operational or practical innovations for landscape transformation that integrates urban agriculture and landscape management goals in a singular, cost-effective framework to support resilience to climate change and better response to challenges in existing urban zones. These challenges include food security, jobs and income-generating activities, and aesthetic, livable, and recreational spaces. Embedding human production in the management of urban landscape and scenic beauty in this way is a useful integrative framework for landscape transformation and management as it can ensure that human production pays for the intangible ecological functions and services (such as scenic beauty) that are public goods while enabling a better response to development demands and climate change. This approach is consistent with recent calls for multifunctionality in agriculture, forestry, and other land use sectors (Russo et al. 2017; Nair and Garrity 2012; Mander et al. 2007).

Inherent in the pilot project was the idea of demonstrating the feasibility of twining human production and maintenance of beautiful landscapes such that the former (as a private good or activity) pays for the latter (as a public good) while supporting urban food security. The human production aspect was fulfilled by the production of high-quality tomato fruits in the greenhouses, with low external inputs such as water and pesticides. The system used in the pilot project produces 2000 kg (2 tonnes) tomato per cycle per greenhouse, and there could be at least three cycles per year. If properly planned with market access to high-end users (such as supermarkets, hotels, and restaurants), three cycles are enough to recover the full cost of installation and operation and generate profit. In this project, eight young men and women were employed for the duration of the project, to work in the greenhouses and maintain the surrounding landscapes. Their initial wages were part of the initial investment cost for the first cycle while payment of wages in subsequent cycles would be paid for by revenue from previous cycles. The initial idea was that these young persons work without wages but receive payment from the postharvest revenues. However, this idea was not acceptable to prospective workers, even those who were already in urban farming. So, in the end, their wages had to be worked into the initial investment cost. The tomatoes produced in the greenhouses were sold locally in the study communities, University of Cape Coast and beyond. Those young persons who worked in the greenhouses also had some tomatoes for their own household consumption. This possibility of paying for ecosystem services, as public good, by human production in urban landscapes to support multiple environmental and development goals was demonstrated during the dissemination event, to stakeholders including representatives from the Cape Coast Metropolitan Assembly, Ministry of Agriculture (regional office), town planning, chiefs and elders of the project communities, tomato retailers association in local markets, academics, and the general public. This approach was intensely discussed during the dissemination event and the consensus was that if this approach is well planned for out- and up-scaling, it can considerably augment urban food security and environmental and human resilience. The members of the project communities indicated how the landscaping could help reduce floods or dust exposure in the communities and improve scenic beauty to bring them some dignity, happiness, and recreational opportunities for children. In all, the project shows a possible pathway for urban employment and livelihoods in food production, landscaping, and landscape management while supporting food security.

Projections suggest that urban food demand will increase substantially due to a combination of increase in population and incomes, increased awareness of food health and safety, and dietary shift (Alexandratos and Bruinsma 2012; Yawson et al. 2020, 2017). A resilient urban community should be food secure and have landscapes that minimize socioecological vulnerability. One way of addressing food safety and security concerns is local supply where the food is transparently produced and freshly supplied to consumers or retailers. Tomato, for example, is almost indispensable in Ghanaian dishes or diets. Tomato production is very seasonal and can be very expensive or simply unavailable during the dry season. Suppliers and market women travel to neighboring countries to bring tomatoes to urban centers in Ghana especially during the dry season. The same can be said of several other vegetables. Yet, there are idle, unkempt spaces in urban areas that currently pose risks and dangers to urban populations and substantially detract from the scenic beauty of urban landscapes. This project demonstrates the fact that a multifunctional edible urban landscape approach can be applied to transform these spaces to productive and livable landscapes to support the multiple goals of food security, jobs and poverty reduction, environmental quality, and human well-being. This approach is not only feasible but also cost-effective and presents a triple-win opportunity for landowners, landscape entrepreneurs and workers, and city managers. It improves landscape structure and ecosystem services to enhance resilience and well-being. In terms of adaptation to climate change, transformation of idle spaces in urban zones to multifunctional edible urban landscapes can be instrumental in strengthening resilience and adaptive capacity in the areas of food insecurity, incomes and livelihoods, and environmental quality.

Lessons and Insights for Scaling Up

Landscaping and maintenance of public outdoor spaces have traditionally been the responsibility of governments. In sub-Saharan Africa, this arrangement seems to be generally absent due to weak financial capability and poor physical planning, resulting in distasteful landscapes that amplify vulnerability to shocks. Although urban agriculture or farming has been practiced for a long time throughout the world, and there is a growing interest and worthy calls for its integration in urban planning (Redwood 2009), it does not necessarily and deliberately integrate landscaping for the purpose of scenic beauty, environmental protection, and recreational use. Similarly, the idea of edible landscapes has been around for a while and practiced in different formats and context, including home or kitchen gardens and agro-parks. However, in Africa where poor financial capability weakens public landscape management, there is scope for exploring the requirements for creating edible landscape entrepreneurs who simultaneously produce food and maintain aesthetic landscapes, with the former paying for the latter. The practical example presented in this chapter tested this idea in parallel with the idea of transforming the idle spaces to multifunctional edible landscape. This project suggested that young persons would not easily defer payment for their work until revenue is generated from sales of produce. Perhaps, they did not consider the project as their own but only as workers, a situation worthy of further exploration. Nonetheless, on a balance, the project result showed that, with careful planning and successful harvest, it is possible to cover this cost from initial revenue and generate profit from the second or third cycle. Hence, the project gives indication that multifunctional edible landscape entrepreneurs can be developed to maintain idle and public spaces that are poorly managed (or not at all) by either the state or private owners due to inadequate resources. Further exploration of this idea would be useful for policy, practice, and research in multifunctional edible urban landscapes to help improve the general landscape character and resilience of urban zones in Africa.

The project used greenhouse system for crop production. This is expensive and could be unaffordable to potential multifunctional edible urban landscape entrepreneurs who would likely be resource-poor. As a result, some form of financial support or business case could be investigated to inform large-scale implementation of this system. While non-greenhouse-based production systems are viable, such as seen in conventional urban farms, the landscape beautification and opportunities for recreational and educational use might give cause for security concerns. Uncontrolled visit or recreational use of the site can expose the crops to unintentional damages. Praedial larceny could also be increased. In addition, because of the quality and safety concerns over vegetables from urban farms in Ghana (Redwood 2009), greenhouse production might be more acceptable to consumers. Finally, because greenhouse production is highly productive, it will not require large area of land to produce profitably and can be more suitable for small idle spaces or where only a small space can be used for food production while the rest of the land area is landscaped. As a result, the use of a greenhouse system as presented in the example project, or otherwise, should be informed by local circumstances, including security concerns.

While Africa is known to be highly vulnerable to climate change, the local-scale impacts of climate change in Africa is poorly understood (Di Ruocco et al. 2015). Cape Coast lies in the coastal savannah agroecological zone and so drought, water scarcity, and heat stress are persistent problems. Cape Coast is the capital of the Central Region, which is the fourth poorest region in Ghana. These challenges would be exacerbated by climate change, population growth, and urbanization. The urban challenges of Cape Coast might not differ significantly from other cities in Ghana. There are numerous opportunities for implementing multifunctional edible urban landscapes in Ghana and sub-Saharan Africa as the characteristics of urban landscapes tend to be similar. The project demonstrated the promise of a multifunctional edible urban landscape approach to strengthening resilience through food production, land cover management, improving urban aesthetics, environmental protection, providing greenspace for recreational and educational opportunities, as well as jobs and livelihood opportunities in cities and towns.

Despite the huge potential of this concept of multifunctional edible urban landscape to contribute to resilience and adaptive capacity, there are success factors that need to be in place and challenges that need to be addressed to ensure success. These are discussed below to help identification of opportunities for innovation and mitigation of challenges and risks that might arise, and guide efforts at implementing or scaling up similar landscape transformation projects in similar contexts.

1) Trust and legitimacy: The first lever of change is developing a strong trust between city authorities (government), landowners, and landscape entrepreneurs. As indicated earlier, the land sector in Ghana is riddled with disputes, protracted litigation, and conflicts (sometimes violent conflicts) due to poor land administration and inordinate pressure on land for housing, commercial, and industrial activities. A parcel of land can be sold to two or more persons by the same or multiple parties. The courts are unable to deal with land disputes or litigation swiftly. Protecting a parcel of land can be more expensive than the cost of the land itself, and sometimes comes at a great cost to parties in a dispute or conflict. In the example project, though earlier verbal assurances were given by landowners to the researchers, the landowners later showed deep distrust and insecurities with releasing their lands for the project. Urban landowners feel insecure granting their lands for temporary projects as there might be limited or costly paths to peaceful recourse to their lands. The first step to successful implementation of this landscape transformation is, therefore, to establish an institutional framework that fosters and deepens a three-way trust between landowners, city authorities, and landscape entrepreneurs. This institutional framework should formalize and render transparent land transactions and use. This is necessary to legitimize the ownership of resources, the process of transformation, formal rights and responsibilities, and benefits sharing. Minimizing insecurities among landowners and strengthening tenure security of landscape entrepreneurs are key to freeing up the numerous idle spaces in existing urban zones for edible urban landscape transformation projects. This, in turn, will require a stronger connection with and improvements in the land administration system at local or national scale to guarantee the security of ownership and access to land by landowners without jeopardizing the tenure security of those who produce on and manage the landscapes.

2) Incentives: Beyond institutional arrangements for land ownership and tenure security, specific incentives can help ease the process of committing idle spaces to edible landscape transformation. As indicated earlier, idle lands do not attract any meaningful taxes and there are no management responsibilities enforced on landowners. There is a need for land management mechanism that encourages owners of idle parcels or spaces to actively manage their lands in a manner that fits the broad urban landscape and ecosystem or transfer such management responsibilities to edible urban landscape initiatives. As part of the institutional arrangements and improvements in land administration, policies and regulatory instruments that impose and enforce land management and tax obligations on landowners would be essential. This needs to be tightly coupled with improvements in land administration structures and procedures to make the process of land registration and titling less cumbersome. In this context, incentives such as free land registration and/or titling, reduced or waving of land taxes, and transfer of land management responsibilities from landowners to landscape entrepreneurs can help catalyze freeing idle lands for multifunctional edible urban landscape initiatives. Neighbors would be supportive of edible urban landscape projects if they become aware of the numerous benefits it can generate to the public, landowners, and the edible landscapists, especially if neighboring residents know that the project would not lead to conflict over land, and that they can benefit from fresh food, recreational and learning opportunities, or aesthetic value of the project sites. The argument of neighborhood landscape improvement could also be made and used to derive small margin of revenue from neighboring properties whose owners or inhabitants could benefit from the aesthetics and the recreational and educational opportunities. A major incentive to landscape entrepreneurs would be a strong connection to markets: local residents, fresh produce retailers, supermarkets, hotels, restaurants, and processors. Local authorities or city managers could facilitate this as part of an integrated approach to serving the multiple goals of urban development demands, including food security, jobs and poverty reduction, and scenic beauty. This will also mean supporting private or individual developers to complete their housing or property development projects quickly to support the achievement of a desirable landscape as a whole.

3) Formal edible landscapes in urban planning: Beyond innovations of landscape transformation in existing urban zones, there is a need to arrest the current chaotic and informal evolution or growth of urban zones and landscapes. This calls for compact urban planning approach that opens up a zone for development at a time, reduces the preponderance of idle spaces, and formally integrates multifunctional edible landscapes in the urban fabric. In the interest of the challenges and threats posed by climate change, resilient cities need to have the capacity for resource efficiency, circular economy, and material transformation or flows. This formal incorporation of multifunctional edible landscapes into the development plan will amount to a formal recognition and practical articulation of the high capacity of urban agriculture for material cycling while supporting multiple socio-ecological services in urban areas (FAO 2012). When greenhouse production is chosen, landscape entrepreneurs might need support as the capital requirement might be high for some. To this end, support for formal access to financial resources and technical support would be helpful and a great incentive to landscape entrepreneurs.

4) Recognition and articulation of landscape needs and services: The first challenge is the need for policymakers, regulators, and urban management authorities to recognize and strongly articulate the fact that the appearance and structure of landscapes in which people live are strongly connected to a sense or state of poverty, powerlessness, vulnerability, despondency, ill-health, and distrust of government (Scott et al. 2009). Urban landscapes can deter or facilitate nefarious or some criminal activities. Urban landscapes can confer confidence, dignity, and a sense of pride (or otherwise) in its inhabitants or users. Recognizing these fundamental relationships between landscapes and people by public authorities is a prerequisite for moving towards policies and management decisions aimed at transforming landscapes to transform lives. People have landscape needs (Matsuoka and Kaplan 2008). These landscape needs are directly linked to the material and existential needs of the landscape inhabitants or users and, by addressing these needs, vulnerability can be reduced, and resilience enhanced (Yawson et al. 2015). A formal recognition and articulation of the linkage between landscapes and urban challenges or people's needs are a prerequisite for innovative planning and transformative approaches that enhance resilience and adaptive capacity. Next to this recognition is the need for political will to tackle the issue of integrated urban planning that can incorporate multifunctional edible urban landscapes, dignity and resilience considerations, or tackle land management in general in urban areas using the approaches or instruments proposed earlier in this chapter. While poverty and undesirable landscapes can have spatial delimitations, their impacts can be diffuse across sectoral, spatial, and temporal scales. The fabric and connectivity of landscapes in urban zones should, therefore, be considered holistically as to ensure broad-based resilience and adaptive capacity across the entire urban space or city. To this end, recognition and articulation of the role of urban landscapes to development and well-being would gain acceptance and support among urban residents and pave the way for mobilizing multiple stakeholders for transformational projects and implementation of planning measures adopted. In Ghana, for example, the Department of Parks and Gardens, which hitherto is responsible for public landscaping and landscape management, seems to be weakened and inactive. Recognizing the link between landscapes and urban resilience and human well-being would also mean revitalizing the appropriate units, such as the Department of Parks and Gardens, Environmental Protection Department, and Agricultural Extension Services, that can provide technical support to landscape entrepreneurs. For example, the Department of Parks and Gardens in Ghana can be a source of genetic or planting materials that helps design the edible landscapes and provides training to prospective landscapists. Context-specific arrangements would be necessary to respond to local challenges or needs. In this way, bottlenecks associated with production and technical assistance in the context of human production and landscape management entrepreneurship can be addressed during planning and operational phases of multifunctional edible landscapes for urban resilience.

Summary and Conclusions

Urban zones in Africa face considerable challenges, including floods, droughts, high rates of poverty, unemployment, food insecurity, and environmental degradation. Climate change, population growth, and urbanization would escalate these challenges. People's landscape needs, poverty, and vulnerability are directly interlinked. This chapter used evidence and insights from a pilot project to promote the feasibility and the instrumental role of a multifunctional edible urban landscape transformation approach that balances land productivity, economic profitability, and human well-being to strengthen resilience and adaptive capacity in poorly planned or managed urban zones in Africa. The pilot project simply tested the idea and workability of incorporating food production into ornamental landscape design, so that the former pays for the latter while maintaining profitability and supporting urban food security, jobs, and livelihoods, and maintenance of scenic beauty and environmental protection to enhance resilience. The findings from the pilot project show that urban spaces which are hitherto idle, unmanaged, and unkempt can be successfully transformed into green, productive, and aesthetically appealing landscapes to support the delivery of multiple ecosystem services and, therefore, contribute to urban resilience. The example project demonstrated the possibility of producing fresh, quality tomato fruits from a small area using a greenhouse, providing jobs or income for young persons, and maintaining ornamentally landscaped surroundings that provide public goods (scenic beauty, safe space for recreational and learning opportunities, and environmental protection). This demonstrates the workability of this landscape transformation approach to resilience and adaptive capacity using spaces which are currently idle and unmanaged in poorly planned and managed urban zones in Africa. The chapter argues that the multifunctional edible urban landscape transformation approach presented in the example project is innovative, cost-effective, and a feasible avenue for green transformation of undesirable landscapes in sub-Saharan Africa to improve lives and well-being. The multifunctional edible urban landscape transformation in this chapter can be adapted and scaled in similar jurisdictions, where limited public funds constrain investment in landscape and scenic beauty management, so that human production (as a private activity) can pay for public goods or ecosystem services. This idea, which requires further exploration, makes multifunctional edible urban landscape transformation, as presented in this chapter, insightful and relevant for policy, practice, and research. The chapter also sheds light on challenges and levers of change critical for successful transition to multifunctional edible urban landscapes as a pathway to enhancing urban resilience in Africa. These include a formal recognition and articulation of the link between urban landscapes and a sense or state of poverty, powerlessness, and vulnerability of inhabitants. This recognition, coupled with political will and a genuine commitment to confronting the problem, can pave the way for building trust and legitimacy for transforming idle spaces, providing incentives for landscape transformation project, and formally incorporating multifunctional edible landscapes in compact urban development planning and processes. Achieving this will, in turn, require multistakeholder mobilization of policies, laws, and institutional arrangements that (i) ease access to idle or formally demarcated spaces in urban zones in a manner that protects or guarantees the security of ownership and tenure for both land owners and landscape entrepreneurs, (ii) incentivizes multifunctional edible urban landscapes in the context of urban planning and management for resilience and adaptation to climate change, (iii) provides support for access to productive and technical resources, and (iv) promote reliable market for produce from multifunctional edible urban landscapes to sustain the human production and the payment for the ecosystem services which are public. Altogether, the landscape transformation approach presented in this chapter demonstrates an innovative path that balances ecological productivity and protection, scenic beauty, and socioeconomic profitability and well-being in existing urban zones, and constitutes a triple-win option for urban managers, landowners, and landscape entrepreneurs. If well planned and effectively managed, it can enhance urban resilience and contribute to long-term sustainability of urban zones in Africa. Hence, multifunctional edible urban landscapes should be integral to the suite of responses to the challenges of climate change.

References

African Union (2020) Framework for irrigation development and agricultural water management in Africa. African Union Commission, Addis Ababa

Alexandratos N, Bruinsma J (2012) World agriculture towards 2030/2050: the 2012 revision. ESA working chapter no. 12-03. Food and Agriculture Organization (FAO) of the United Nations, Rome

Armah FA, Odoi JO, Yengoh GT, Obiri S, Yawson DO, Afrifa EKA (2011) Food security and climate change in drought-sensitive savanna zones of Ghana. Mitig Adapt Strateg Glob Chang 16:291–306

Armah FA, Ekumah B, Yawson DO, Odoi JO, Afitiri A-R, Nyieku FE (2018) Access to improved water and sanitation in sub-Saharan Africa in a quarter century. Heliyon 4:e00931. https://doi.org/10.1016/j.heliyon.2018.e00931

Barton J, Pretty J (2010) What is the best dose of nature and green exercise for improving mental health? A multi-study analysis. Environ Sci Technol 44(10):3947–3955

Batchelor C, Schnetzer J (2018) Compendium on climate-smart irrigation: concepts, evidence and options for a climate-smart approach to improving the performance of irrigated cropping systems. Global Alliance for ClimateSmart Agriculture, Rome

Bratman GN, Hamilton JP, Daily GC (2012) The impacts of nature experience on human cognitive function and mental health. Ann N Y Acad Sci 1249:118–136. https://doi.org/10.1111/j.1749-6632.2011.06400.x

Bryce J, Boschi-Pinto C, Shibuya K, Black RE (2005) The WHO child health epidemiology reference group. WHO estimates of the causes of death in children. Lancet 365:1147–1152

Bustamante M, Robledo-Abad C, Harper R, Mbow C, Ravindranath NH, Sperling F, Haberl H, Pinto AD, Smith P (2014) Co-benefits, trade-offs, barriers and policies for greenhouse gas mitigation in the agriculture, forestry and other land use (AFOLU) sector. Glob Chang Biol 20: 3270–3290

Chapman S, Watson JEM, Salazar A, Thatcher M, McAlpine CA (2017) The impact of urbanization and climate change on urban temperatures: a systematic review. Landsc Ecol 32:1921–1935. https://doi.org/10.1007/s10980-017-0561-4

Coutts C, Horner M, Chapin T (2010) Using geographical information system to model the effects of green space accessibility on mortality in Florida. Geocarto Int 25(6):471–484

Csortan G, Ward J, Roetman P (2020) Productivity, resource efficiency and financial savings: an investigation of the current capabilities and potential of South Australian home food gardens. PLoS One 15(4):e0230232. https://doi.org/10.1371/journal.pone.0230232

Dai D (2011) Racial/ethnic and socioeconomic disparities in urban green space accessibility: where to intervene? Landsc Urban Plan 102(4):234–244

Di Ruocco A, Gasparini P, Weets G (2015) Urbanization and climate change in Africa: setting the scene. In: Pauleit S et al (eds) Urban vulnerability and climate change in Africa. Future city, vol 4. Springer, Geneva, pp 1–35. https://doi.org/10.1007/978-3-319-03982-4_1

Eschobedo FJ, Kroeger T, Wagner JE (2011) Urban forests and pollution mitigation: analyzing ecosystem services and disservices. Environ Pollut 159(8):2078–2087

FAO (2012) Growing greener cities in Africa. First status report on urban and peri-urban horticulture in Africa. Food and Agriculture Organization (FAO) of the United Nations, Rome

FAO, ECA, AUC (2020) Africa regional overview of food security and nutrition 2019. Food and Agriculture Organization (FAO) of the United Nations, Accra. https://doi.org/10.4060/CA7343EN

Fischer AP, Vance-Borland K, Jasny L, Grimm KE, Charnley S (2016) A network approach to assessing social capacity for landscape planning: the case of fire-prone forests in Oregon, USA. Landsc Urban Plan 147:18–27

Ghana Statistical Service (2013) 2010 population and housing census: District analytical report - Cape Coast Municipality. Ghana Statistical Service, Accra, Ghana

Grêt-Regamey A, Weibel B, Kienast F, Rabe S-E, Zulian G (2015) A tiered approach for mapping ecosystem services. Ecosyst Serv 13:16–27

Güneralp B, Lwasa S, Masundire H, Parnell S, Seto KC (2017) Urbanization in Africa: challenges and opportunities for conservation. Environ Res Lett 13:015002

Haase D, Larondelle N, Andersson E et al (2014) A quantitative review of urban ecosystem service assessments: concepts, models, and implementation. Ambio 43(4):413–433. https://doi.org/10.1007/s13280-014-0504-0

Haberman D, Gillies L, Canter A, Rinner V, Pancrazi L, Martellozzo F (2014) The potential of urban agriculture in Montréal: a quantitative assessment. ISPRS Int J Geo Inf 3(3):1101–1117. https://doi.org/10.3390/ijgi3031101

Health Effects Institute (2019) State of global air 2019. http://www.stateofglobalair.org. Accessed 18 Sep 2019

IPCC (2007) Climate change 2007: impacts, adaptation and vulnerability. Contribution of working group II to the fourth assessment report of the Intergovernmental Panel on Climate Change (IPCC). Cambridge University Press, Cambridge, UK

Khan RK, Strand MA (2018) Road dust and its effect on human health: a literature review. Epidemiol Health 40:e2018013. https://doi.org/10.4178/epih.e2018013

Laille P, Provendier D, Colson F (2014) The benefits of urban vegetation: a study of the scientific research and method of analysis. Plante & Cité, February 2014, 28 p. https://www.plante-et-cite.fr/data/beneveg_english_bd.pdf. Accessed June 22, 2020

Larbi WO (2008) Compulsory land acquisition and compensation in Ghana: searching for alternative policies and strategies. FIG/FAO/CNG International Seminar on State and Public Sector Land Management, Verona, Italy, September 9–10, 2008, 21 p. http://www.fig.net/commission7/verona_fao_2008/chapters/09_sept/4_1_larbi.pdf. Accessed 4 Dec 2013

Larbi WO, Antwi A, Olomolaiye P (2004) Compulsory land acquisition in Ghana: policy and praxis. Land Use Policy 1(2):115–127

De Longueville F, Hountondji Y-C, Henry S, Ozer P (2010). What do we know about effects of desert dust on air quality and human health in West Africa compared to other regions? Sci Total Environ 409(1):1–8

Lwasa S (2014) Managing African urbanization in the context of environmental change. Interdisciplina 2(2):263–280

Mander Ü, Helming K, Wiggering H (eds) (2007) Multi-functional land use: meeting future demands for landscape goods and services. Springer-Verlag Berlin Heidelberg, Berlin, 13 p. Ch. 1

Masten AS (2014) Ordinary magic: resilience in development. Guilford Press, New York

Matsuoka RH, Kaplan R (2008) People needs in the urban landscape: analysis of Landscape and Urban Planning contributions. Landsc Urban Plan 84(1):7–19

Mensah CA (2014) Urban green spaces in Africa: nature and challenges. Int J Ecosyst 4(1):1–11

Morris SS, Black RE, Tomaskovic L (2003) Predicting the distribution of under-five deaths by cause in countries without adequate vital registration systems. Int J Epidemiol 32:1041–1051

Nair PKR, Garrity D (eds) (2012) Agroforestry – the future of global land use. Springer, Dordrecht

Nowak DJ, Crane DE, Stevens JC (2006) Air pollution removal by urban trees and shrubs in the United States. Urban Forest Urban Green 4:115–123

Panagopoulos T, Duque JAG, Dan MB (2016) Urban planning with respect to environmental quality and human well-being. Environ Pollut 208:137–144

Redwood M (ed) (2009) Agriculture in urban planning: generating livelihoods and food security. Earthscan and IDRC, Sterling

Russo A, Escobedo FJ, Cirella GT, Zerbe S (2017) Edible green infrastructure: an approach and review of provisioning ecosystem services and disservices in urban environments. Agric Ecosyst Environ 242:53–66

Santika T, Meijaard E, Wilson KA (2015) Designing multi-functional landscapes for forest conservation. Environ Res Lett 10:114012

Säumel I, Reddy SE, Wachtel T (2019) Edible city solutions – one step further to foster social resilience through enhanced socio-cultural ecosystem services in cities. Sustainability 11(4):972

Scott AJ, Shorten J, Owen R, Owen I (2009) What kind of countryside do the public want: community visions from Wales UK? GeoJournal. https://doi.org/10.1007/s10708-009-9256-y

Stern N, Treasury G (2007) The economics of climate change: the Stern review. Cambridge University Press, Cambridge, UK

Tham-Agyekum EK, Okorley EL, Amamoo FA (2019) Alternative livelihood support for reducing poverty: snail project for Kwaprow Community in Cape Coast. J Advocacy Res Educ 6(2):33–39

Thompson CW (2011) Linking landscape and health: the recurring theme. Landsc Urban Plan 99: 187–195

UN DESA (2012) World urbanization prospects: the 2011 revision. United Nations Department of Economic and Social Affairs, Population Division, United Nations, New York. http://esa.un.org/unup/CD-ROM/Urban-rural-Population.htm. May 29, 2017

UN DESA (2019) World urbanization prospects: the 2018 revision. ST/ESA/SER.A/420. United Nations, Department of Economic and Social Affairs, Population Division, New York

Urban Agriculture (2009) Building resilient cities. Urban Agriculture Magazine 22, June 2009, 56 p

van Noordwijk M, Agus F, Dewi S, Purnomi H (2014) Reducing emissions from land use in Indonesia: motivation, policy instruments and expected funding streams. Mitig Adapt Strateg Glob Chang 19:677–692

Van Rooyen AF, Ramshaw P, Moyo M, Stirzaker R, Bjornlund H (2017) Theory and application of Agricultural Innovation Platforms for improved irrigation scheme management in Southern Africa. Int J Water Res Dev 33(5):804–823

Villeneuve PJ, Jerret M, Su JG et al (2012) A cohort study relating urban green space with mortality in Ontario, Canada. Environ Res 115:51–58

WHO (2018) World malaria report 2018. World Health Organization, Geneva. https://apps.who.int/iris/bitstream/handle/10665/275867/9789241565653-eng.pdf?ua=1. Accessed 1 July 2020

Wolch JR, Byrne J, Newell JP (2014) Urban green space, public health, and environmental justice: the challenge of making cities 'just green enough'. Landsc Urban Plan 125:234–244

Yawson DO (2020) Quantifying perceived landscape desirability in human settlements: the case of four communities in Cape Coast, Ghana. Ghana J Geogr 12(1):74–98

Yawson DO, Armah FA (2018) Alienation and conflicts in protected forest resource exploitation in Ghana: who or what is legal? Proceedings of the Cave Hill Philosophy Symposium, Conversation XIII, The University of the West Indies, Barbados 13(1), 22 p

Yawson DO, Adu MO, Armah FA, Kusi J, Ansah IG, Chiroro C (2015) A needs-based approach for exploring vulnerability and response to disaster risk in rural communities in low income countries. Australas J Disaster Trauma Stud 19:27–36

Yawson DO, Mulholland BJ, Ball T, Adu MO, Mohan S, White PJ (2017) Effect of climate and agricultural land use changes on UK feed barley production and food security to the 2050s. Land 6(4):74. 14pp

Yawson DO, Adu MO, Asare PA, Armah FA (2019) Prospects and challenges of edible urban landscapes in Ghana. An AgriFose2030 research brief, University of Cape Coast (Ghana) and University of Gothenburg (Sweden)

Yawson DO, Mohan S, Armah FA, Ball T, Mulholland B, Adu MO, White PJ (2020) Virtual water flows under projected climate, land use and population change: the case of UK feed barley and meat. Heliyon 6:e03127

8

Economic Analysis of Climate-Smart Agriculture Technologies in Maize Production in Smallholder Farming Systems

Angeline Mujeyi and Maxwell Mudhara

Contents

Introduction .
CSA in Crop-Livestock Farming Systems .
Methodology
    Study Area and Data Collection
    Data Analysis
Results and Discussion
    CSA Adaptation Strategies Employed by Households in Maize Production
    Economic Analysis of Maize ....
    Estimated Stochastic Frontier Profit Function
Conclusions and Recommendations
References

Abstract

Smallholder farmers who grow the staple maize crop rely mainly on rain-fed agricultural production, and yields are estimated to have decreased by over 50% largely due to climate change. Climate-smart agriculture (CSA) technologies, as adaptive strategies, are thus increasingly being promoted to overcome problems of declining agricultural productivity and reduced technical efficiency. This study analyzed profitability and profit efficiency in maize (Zea mays) production as a result of CSA technology adoption using cost-benefit analysis and stochastic profit frontier model. The study used data from a

A. Mujeyi (✉) · M. Mudhara
College of Agricultural, Engineering and Science, Discipline of Agricultural Economics, University of KwaZulu-Natal, Scottsville, Pietermaritzburg, South Africa
e-mail: Mudhara@ukzn.ac.za

cross-sectional household survey of 386 households drawn from 4 districts in Mashonaland *East* province located in the northeastern part of *Zimbabwe*. Results from the cost-benefit analysis reveal that maize performs best under CSA technologies. The profit inefficiency model shows that extension contact, number of local traders, and adoption of CSA had significant negative coefficients indicating that as these variables increase, profit efficiency among maize-growing farmers increases. This implies that profit inefficiency in maize production can be reduced significantly with improvement in extension contact, access to farm gate/local markets, and adoption of CSA. The findings call for development practitioners to incorporate market linkages that bring buyers closer to the farmers, support for extension to be able to have frequent contacts with farmers, and promotion of CSA adoption.

Keywords

Cost-benefit analysis · Return on investment · Profit efficiency · Stochastic frontier · Zimbabwe

Introduction

Maize (*Zea mays* L.) is the most important cereal crop in sub-Saharan Africa and is the world's most widely grown cereal crop as well as essential food source for millions of the world's poor (Abate et al. 2017). In sub-Saharan Africa, maize is a staple food for an estimated 50% of the population and an important source of carbohydrate, protein, iron, vitamins (A, B, E, and K), and minerals (magnesium, potassium, and phosphorus) and is grown on an estimated 100 million hectares throughout the developing world (Nsikak-Abasi and Okon 2013; Siyuan et al. 2018). In 2018, Zimbabwe got approximately 730,437 tonnes of maize, and the average yield was 613.1 Kg per hectare pointing to some technical inefficiencies. The average yield is lower as compared to the world average of 5923.7 Kg tonnes/ha and 2040.2 Kg/ha for Africa in the same year (FAOSTAT 2020). The smallholder farmers rely mainly on rain-fed production and in addition are often constrained by multiple constraints such as reduced soil fertility; limited income to access inputs such as fertilizers, improved seed, herbicides, and pesticides; unavailability of lucrative output markets; high cost of inputs; and reduced yield due to climate variability (Poole 2017; Rurinda et al. 2014). Researchers and development practitioners have reported reductions in agricultural yield due to extreme weather (UNCCS 2019). These unpredictable seasons have become a major constraint in smallholder crop and livestock production farming systems, and as such, the use of climate-smart agriculture (CSA) technologies becomes essential as a solution. Climate-smart agriculture technologies are innovations that sustainably increase agricultural productivity, help households to adapt and be resilient to climate change, and contribute to the reduction of greenhouse gas emissions (Steward et al. 2018).

Adaptation strategies for households can either be reactive (Shongwe et al. 2014), i.e., soil fertility maintenance through the use of animal manure and inorganic fertilizers, rotations, and intercropping in order to address problems linked to observed climate change impacts, or proactive CSA technologies such as use of new drought-tolerant varieties, use of early maturing varieties, and policy measures such as insurance policy. Zimbabwe has participated in interventions and alliances promoting CSA such as the DFID-funded Vuna (2015–2018) and the Africa Development Bank's Africa Climate-Smart Agriculture (ACSA) (2018–2025) (Thierfelder et al. 2017; Rosenstock et al. 2019). The Government of Zimbabwe (GoZ) has developed policies and interventions to lessen the impacts of climate change on agriculture. These policies include a child-friendly climate policy which targets education in schools on climate change issues, the climate-smart agriculture policy which promotes adoption of CSA by farmers, and the national climate policy which is targeting putting legal structures to guide businesses on becoming greener (GOZ 2018). Government and nongovernmental organizations have introduced a range of CSA in Zimbabwe which include conservation agriculture, drought-tolerant maize and legume varieties, cereal-legume intercropping and rotation systems, and improved fodder crops among others (Mujeyi 2018). Assuming economic rationality, smallholder farmers who rely on agriculture for livelihoods would adopt technologies that reduce costs of production while increasing benefits from greater incomes through improved yields. Smallholder farmers are heterogeneous, and as such, they adopt different combinations of CSA to address varying constraints that they face. These different technology bundles have different profitability levels because of the different input requirements associated with them as well as their potential impact on productivity.

The need to upscale CSA as adaptation mechanisms in order to improve or maintain high productivity levels in smallholder farming communities can effectively be achieved if profitability of these technologies and factors that enhance efficiency are properly understood. This study therefore aims to:

1. Estimate profitability and compare benefit-cost ratio (BCR) of maize production in smallholder farming communities across CSA technology bundles
2. Investigate the determinants of profit efficiency and identify the determinants thereof

The aim of this study is to contribute to the literature on CSA in Zimbabwe by analyzing profitability of current CSA technology bundles in maize production and technical inefficiency. Furthermore, using stochastic frontier model, the chapter aims to identify determinants of efficiency. The results will provide a better understanding of costs and benefits that would make it possible to design more economically efficient policies and programs to promote CSA technology adoption. Economic evaluations can provide critical information to those making decisions about the allocation of limited agriculture input resources across enterprises. The chapter provides empirical evidence from actual farmer behavior in uncontrolled environment, thus adding to studies from on-farm and on-station trials.

CSA in Crop-Livestock Farming Systems

This study particularly chose to do analysis for maize (*Zea mays*) as it is the most important crop in smallholder farming systems in the four districts. Maize is the staple crop in Zimbabwe to 98% of the 12.7 million people in the country, and it provides 40–50% of the calories (Kassie et al. 2017). Average maize yield has dropped from a highest (after independence) of 2163.7 Kg/ha in 1985 to 667.8 Kg/ha in 2017 (FAOSTAT 2020). Maize productivity has been negatively affected by infertile soils, inadequate water due to drought, and erratic rainfall patterns caused by climate change as well as incidence of pests and diseases. Various CSA technologies have been used in maize production in an effort to boost yields. One such technology is conservation agriculture (CA) which consists of three key principles, namely, minimum tillage, permanent soil cover (mulching with crop residues or cover crop), and crop diversification (either temporal diversification, i. e., rotation, or spatial diversification, i.e., intercropping). CA offers benefits of increased yields when properly followed. Crop rotation and intercropping improve soil fertility through the nitrogen fixing characteristics of legumes. Large increases in maize yields in maize-groundnut rotations have been reported by CIMMYT researchers in Zimbabwe from long-term trials in smallholder farming systems (Waddington et al. 2007). Cereal-legume rotations also have benefits of reducing build-up of pests and diseases. Minimum soil disturbance reduces the rate and amount of soil erosion. Soil cover leads to reduced runoff, reduced soil erosion, increased water infiltration, and reduced evaporation of soil moisture (Michler et al. 2019; Steward et al. 2018; Thierfelder et al. 2017). Drought-tolerant maize (DTM) varieties have been promoted by organizations such as CIMMYT, and these are input-responsive, stress-tolerant, and high-yielding in comparison to traditionally grown commercial hybrids (Mujeyi and Mujeyi 2018).

Methodology

Study Area and Data Collection

This study uses data collected from a cross-sectional household survey using a structured interview in communities that were exposed to CSA technologies and data from key informant interviews with stakeholders who were involved in technology dissemination. Multistage sampling method was used to select the 386 respondents from maize-growing communities in 8 wards located in 4 districts, i.e., Goromonzi, Murehwa, Uzumba Maramba Pfungwe, and Mutoko. The economies of the four districts are integrated crop-livestock farming systems that rely on rain-fed production. Maize is the main cereal staple crop, while groundnut is the leading legume crop. The main livestock kept by the farmers are cattle and goats.

Livestock rely mainly on pastures for feed. Integration of the crop and livestock enterprises helps farmers to maximize resource uses. Stover from the field crops are used to feed livestock, while dung from the livestock is used to improve soil fertility through its use as manure.

Murehwa district falls under agro-ecological region IIB which is characterized by moderately high rainfall (700 mm annually) and moderate temperatures for crop production. This district has predominantly sandy loamy soils. The majority of Motoko's communal area is in natural region IV which is characterized by subtropical climate with cool dry winter and hot rainy summers (650 to 700 mm rainfall annually). The soils are shallow to moderately deep, yellowish red, coarse-grained loamy sands. Goromonzi is located in natural region II which also gets moderately high rainfall. Uzumba Maramba Pfungwe (UMP) has two natural regions (natural regions II to V), but wards were selected from natural region V.

Two wards that have been exposed to CSA technologies were chosen from each district. Households were randomly selected from one randomly selected village in each ward. Sample households were distributed within the wards according to the ward sizes (proportionate sampling). The farm households were interviewed by trained enumerators during the 2017/2018 crop season.

Data Analysis

The study employed descriptive statistics and inferential statistics. It explored the economic assessment of CSA technologies through a cost-benefit analysis (CBA) and a stochastic profit frontier model. This study precisely probed farmers to state which CSA technologies they had used for various crops in one season and the inputs that were used and grain harvested after such an investment. Information from this economic analysis is important for price setting of commodities by government watchdogs, researchers working to improve the technologies, farmers using them, and donors and governments who fund research and development work.

Economic Analysis of CSA

Farmers use different technologies as adaptation strategies, and their decisions on which technology to adopt under what area depend on the cost-effectiveness (Shongwe et al. 2014). Cost-benefit analysis thus plays an important role of farmers' decisions with regard to input costs, e.g., fertilizer, labor, seed, pesticide, etc., and was used in the economic analysis. Other researchers have used CBA in analyzing CSA technologies (Papendiek et al. 2016; Sain et al. 2017). Cost-benefit analysis (CBA) compares inputs and outputs for a technology in monetary terms (Shongwe et al. 2014). CBA for this study focuses on the quantitative evaluation of CSA technologies on the maize crop. All benefits and costs are estimated in monetary terms, and through calculating net benefits, the most economic efficient CSA are identified. Benefits from maize include grain and stover used to feed livestock. The net benefits are calculated as follows:

$$NB = \Sigma \, (Bt - Ct) \tag{1}$$

$$NB = \Sigma Bt - \Sigma Ct \tag{2}$$

where:

NB represents the net benefits.

ΣBt = total benefits in year t.

ΣCt = total variable costs (TVC) in year t.

Bt is the combination of revenue from quantity of grain output and stover benefits.

$$\Sigma Bt = \text{Total Revenue}$$
$$= \Sigma \, (\text{Grain Output (Kg)} * \text{Unit grain prices (\$/Kg)}) \tag{3}$$
$$+ (\text{Stover Output (Kg)} * \text{Unit stover prices})$$

Average local market prices obtained by the farmers were used to compute returns. The farm gate price of the output is the value (price) farmers receive or can receive for their harvested crops. Total variable input costs refer to the sum of all variable input costs and vary from one CSA technology to another.

$$TVC = \Sigma Ct$$
$$= P_{\text{landprep}}Q_{\text{landprepr}} + P_{\text{basalfertiliser}}Q_{\text{basalfertiliser}}$$
$$+ P_{\text{topdressingfertiliser}}Q_{\text{topdressingfertiliser}} + P_{\text{seed}}Q_{\text{seed}} + P_{\text{labor}}Q_{\text{labor}} + \cdots$$
$$+ P_n Q_n \tag{4}$$

The benefit-to-cost ratio (BCR) which is a financial ratio that is used to determine whether the amount of money made through a project will be greater than the costs incurred in executing was also computed as follows:

$$BCR = (\text{Benefit/Costs}) \tag{5}$$

For each CSA technology, the total costs incurred when using that strategy and benefits were used to compute the net benefit for that particular adaptation strategy.

Return on Investment

Return on investment values help link the value of technologies to users. The return on investment (ROI) value is more powerful than the benefit-cost ratio because the ROI value shows the net return for a $100 investment.

$$ROI = (\text{Net Benefit/TVC}) * 100 \tag{6}$$

The Stochastic Profit Frontier Model

The stochastic frontier models have been used extensively even in agriculture, to model input-output relationships and to measure the technical efficiency (Greene 2010). These were first proposed in the context of production function estimation to account for the effect of technical inefficiency (Wang 2008; Dziwornu and Sarpong 2014). The analytical method has been used to compare the performance of farmers under different technological regimes. For example, the method has been used to examine the impact of technology adoption on output and technical efficiency of rice farmers or even beef farmers under various production systems (Omhile et al. 2016; Villano and Fleming 2006). In this study, the stochastic profit frontier model is used to compare inefficiency of farmers using CSA versus those who are not using any CSA technology. The model captures inefficiencies associated with different endowments as well as input and output prices. The model is specified as follows:

$$y = \beta'x + \varepsilon_i \tag{7}$$

where y is the observed outcome in this case maize profitability estimated by the gross margin (goal attainment), x is the logarithm of costs of that input, coefficient β are parameters estimated, and ε_i is the error term. The error structure is specified as follows:

$$\varepsilon_i == v_j - u_j \tag{8}$$

where v_j is the random error term and u_j is the inefficiency effects of farm j.
    Uj is independently distributed with mean μ_1 and variance σ^2_u.
    Thus, the stochastic model is:

$$y = \beta'x + v_j - u_j, \tag{9}$$

$\beta'x + v$ is the optimal, frontier goal (e.g., maximal production output or minimum cost) pursued by the individual, $\beta'x$ is the deterministic part of the frontier, and $v \sim N$ $[0, \sigma v^2]$ is the stochastic part. v_j is the stochastic error term, and u_j is a one-sided error representing the technical inefficiency of firm j. Both v_j and u_j are assumed to be independently and identically distributed.
    Inefficiency model is modelled using farm-specific, market-specific, and household characteristics and can therefore be estimated as follows:

$$U_j = \alpha + \alpha_i Z_i + \varepsilon i \tag{10}$$

$$U_j = \alpha + \alpha_1 Z_{1} + \alpha_2 Z\alpha_{2} + \alpha_3 Z_3 + \ldots \alpha_n Z_n + \varepsilon_i \tag{11}$$

where U_j is technical inefficiency of the jth farm.
    Z_1 to Z_n are the determinants and ε_i is the disturbance term and the coefficients $\propto$ are parameters estimated. Stochastic frontier models allow to analyze technical inefficiency in the framework of production functions. Production units such as

households are assumed to produce according to a common technology and reach the frontier when they produce the maximum possible output for a given set of inputs. Inefficiencies can be due to structural problems or market imperfections and other factors which cause countries to produce below their maximum attainable output. The stochastic frontier model decomposes growth of the output variable into changes in input use, changes in technology, and changes in efficiency. All parameters in the stochastic frontier and the technical inefficiency effects model are simultaneously calculated by a single-stage maximum likelihood estimation procedure using sfcross command in Stata (Karakaplan 2017). Table 1 gives a summary of all the variables thus used in the stochastic frontier model.

Table 1 Stochastic frontier model variables

Frontier regression model (efficiency factors)			
yi	yi	Dependent variable – maize gross margin in US$	Continuous variable
X1	SEEDcosts	Seed costs in US$	Continuous variable
X2	DFERTcosts	Basal fertilizer costs in US$	Continuous variable
X3	ANFERTcosts	Top dressing fertilizer costs in US$	Continuous variable
X4	LANDPREPcosts	Land preparation costs in US$	Continuous variable
X5	MANUREcosts	Manure costs in US$	Continuous variable
X6	HERBcosts	Herbicide costs in US$	Continuous variable
X7	PESTcosts	Pesticide costs in US$	Continuous variable
X8	LABOURcosts	Labor costs in US$	Continuous variable
X9	PACKcosts	Packaging costs in US$	Continuous variable
X10	OTHERcosts	Other costs in US$	Continuous variable
Inefficiency model			
Z1	HHSEX	Gender of household head	Dummy, i.e., 1 = male 0 = female
Z2	HHEXPER	Experience household head (years)	Continuous variable
Z3	MEMBERSHIP	Membership to farmer groups	Dummy, i.e., 1 = yes 0 = no
Z4	CREDIT	Access to credit	Dummy, i.e., 1 = yes 0 = no
Z5	TRADERS	Number of traders locally	Continuous variable
Z6	TAR	Distance to tar (km)	Continuous variable
Z7	Kmextension	Distance to extension (Km)	Continuous variable
Z8	TLU	Total livestock units	Continuous variable
Z9	AGROREGION	Agro-ecological region	Dummy, i.e., 1 = wetter (II) 0 = otherwise (drier III and IV)
Z10	EXTNcontact	Frequency of extension contact	Continuous variable
Z11	CSAadoption	Use of CSA in maize production	Dummy, i.e., 1 = yes 0 = otherwise

Table 2 Maize CSA technologies

Maize technology	Goromonzi	Murehwa	Mutoko	U.M.P	Whole sample	Chi square
Intercropping	24.0%	21.6%	2.0%	5.4%	16.1%	24.23***
Sole CN	5.5%	7.2%	0.0%	6.5%	5.4%	3.66
Rotation	39.0%	54.6%	66.0%	47.3%	48.4%	12.88**
Minimum tillage	39.0%	35.1%	48.0%	24.7%	35.8%	8.89**
DT maize	13.7%	11.3%	36.0%	12.9%	15.8%	17.85***
Manure use	13.7%	21.6%	14.0%	8.6%	14.5%	6.69*
Mulching	4.1%	5.2%	10.0%	0.0%	4.1%	8.59**

***, **, and * indicates significance level at 1%, 5% and 10%

Results and Discussion

Profitability across CSA technology bundles was estimated using cost-benefit analysis, and the stochastic profit frontier model was estimated to see if CSA adoption has a significant effect on technical inefficiency. Tables 2, 3, and 4 show the results of the analysis with subsequent discussions.

CSA Adaptation Strategies Employed by Households in Maize Production

Maize production is negatively affected by climate change, and as such, adoption of CSA technologies is key to increasing yields. Table 2 shows the CSA technologies currently being used by the farmers.

The results show that farmers use various CSA technologies in maize production, with crop rotation being the highest in Mutoko followed by Murehwa (66% and 54.6%, respectively). Minimum tillage and DT maize are highest in Mutoko (48% and 36%, respectively). Few farmers (less than 10%) are not using any CSA technologies in maize production. This highlights the importance of CSA in the smallholder farming communities. Adoption of CSA such as intercropping, rotation, minimum tillage, DT maize, manure use, and mulching was significantly different across the study districts. Overall, CSA technology use is still low with less than 50% of households adopting CSA across all the districts except for rotation which is adopted by more than 50% of households in Murehwa and Mutoko districts. Farmers highlighted during FGDs that manure use had become low as there was an outbreak of theileriosis which led to most households being left with no cattle, which are the major source of manure. Manure from small ruminants and poultry is prioritized for use in horticulture gardens. Farmers also cited that technologies such as minimum tillage promoted by NGOs particularly basin making with hoes were strenuous in as much as they could be done bit by bit in the dry season for farmers with fencing. This was not so for the majority with unfenced fields who therefore needed to do it at the onset of the season. This has led to farmers shunning basins in favor of even hiring in

Table 3 Results of cost-benefit analysis

Cost-benefit indicators	Maize technology cluster				
	Cluster 1 N = 178	Cluster 2 n = 163	Cluster 3 n = 24	Cluster 4 n = 21	ALL n = 386
Grain (Kg)	1646.41	1815.61	1833.51	1266.87	1711.02
Grain revenue ($)	643.94	705.14	752.63	488.18	668.91
Stover (Kg)	823.21	907.80	916.75	633.43	855.51
Stover revenue ($)	32.93	36.31	36.67	25.34	34.22
Total revenue	**676.87**	**741.45**	**789.30**	**513.52**	**703.13**
Land preparation costs	68.85	65.37	67.81	77.46	67.75
Seed (Kg)	25.72	25.20	26.60	29.76	25.78
Seed costs ($)	67.60	71.71	69.56	68.73	69.59
Compound D fertilizer (Kg)	204.97	208.33	247.40	180.58	207.80
Compound D fertilizer costs	137.76	138.44	151.12	134.94	138.76
Ammonium nitrate fertilizer (Kg)	184.39	187.66	192.53	178.17	185.99
Ammonium nitrate fertilizer costs ($)	137.43	137.46	137.08	141.96	137.68
Manure (carts)	0.00	0.02	0.00	0.00	0.01
Manure costs ($)	30.39	33.22	47.60	30.16	32.72
Herbicide costs ($)	1.55	2.01	0.29	0.48	1.61
Pesticide costs ($)	0.38	0.23	2.08	0.00	0.40
Labor costs ($)	66.36	72.74	47.67	119.05	70.91
Maize packaging costs ($)	5.02	6.68	5.05	4.33	5.71
Other costs ($)	0.21	0.88	2.03	0.00	0.61
Total variable costs (TVC)	**515.56**	**528.75**	**530.30**	**577.11**	**525.74**
Gross margin	**161.30**	**212.70**	**259.00**	**−63.59**	**177.39**
BCR	**1.42**	**1.50**	**1.69**	**0.90**	**1.44**
ROI	**42.17**	**50.06**	**68.82**	**−9.59**	**44.42**

***, **, and * indicates significance level at 1%, 5% and 10%

animal-based tillage services. Minimum tillage could be achieved using animal-drawn rippers and direct seeders, but farmers highlighted that there has been an outbreak of January diseases which saw farmers losing cattle and draft power was the hardest hit. Mulching and intercropping under maize also recorded the least frequencies. Farmers highlighted that mulching was difficult to come by given that stover was used to feed livestock. The study further identified CSA technology combinations in maize production using principal component analysis-clustering. Four distinctive clusters were identified, i.e., Technology Cluster 1 (dominantly minimum tillage with lower use of rotation, DT maize, manure, and intercrop), Technology Cluster 2 (dominantly rotation use with lower use of intercrop and very low DT, manure, and minimum tillage), Technology Cluster 3 (higher use of mulch, manure, and DT maize, average use of minimum tillage and rotation, and less intercrop), and Technology Cluster 4 (conventional).

Table 4 The stochastic frontier model results

Variables		Coef.	Std. Err	P value
Frontier regression model (efficiency factors)				
X1	SEEDcosts	102.41	151.51	0.50
X2	DFERTcosts	−166.68**	67.53	0.01
X3	ANFERTcosts	−40.02	67.85	0.56
X4	LANDPREPcosts	106.57	105.16	0.31
X5	MANUREcosts	11.6	27.17	0.67
X6	HERBcosts	−93.47	74.80	0.21
X7	PESTcosts	15.98	121.90	0.90
X8	LABOURcosts	−28.28	24.82	0.25
X9	PACKcosts	1362.15***	79.66	0.00
X10	OTHERcosts	−208.07**	96.51	0.03
	_cons	−642.06	324.50	0.05
Inefficiency model				
Z1	HHSEX	−51.86	72.02	0.47
Z2	HHEXPER	152.62**	69.20	0.03
Z3	MEMBERSHIP	18.08	63.86	0.78
Z4	CREDIT	117.29	76.06	0.12
Z5	TRADERS	−145.16**	60.61	0.02
Z6	TAR	−74.88	85.25	0.38
Z7	Kmextension	100.71	64.87	0.12
Z8	TLU	181.94***	59.94	0.00
Z9	AGROREGION	−60.21	63.55	0.34
Z10	EXTNcontact	−167.5**	82.10	0.04
Z11	CSAadoption	−297.64**	125.80	0.02
	_cons	436.91**	201.05	0.03
Usigma				
_cons		4.65	7.77	0.55
Vsigma				
_cons		11.78***	0.09	0.00
sigma_u		10.22	39.70	0.80
sigma_v		361.46***	15.36	0.00
Lambda		0.03	42.72	1.00

***, **, and * indicate statistical significance at 1%, 5%, and 10%, respectively

Economic Analysis of Maize

Economic analysis was performed to estimate the net return and benefit-cost ratio in various CSA technology bundles. A comparison of costs and returns from various CSA technology combinations in maize production is presented in Table 3.

The results show that the farmers who used CSA had higher gross margin ranging from $259 (return on investment of 69%) with a BCR of 1.69 under higher CSA use to \$161.30 (return on investment of 42%) and a BCR of 1.42 under low CSA use compared to a negative gross margin under sole conventional practices (−\$63.59) with a BCR of 0.9 but negative ROI of close to 10%. This indicates that farmers get at least more than \$40 for every \$1 spent in maize production using CSA technologies. The difference in profitability is mainly maybe a result of yield differences of conventional system versus CSA. These findings are consistent with the findings of Sain et al. (2017) who found that the incorporation of the CSA practices increased maize yields by 20% or more in comparison to existing farm management systems and Ali and Erenstein (2017) who found that yields differed according to production system and technology used.

Estimated Stochastic Frontier Profit Function

The analysis was done using the sfcross Stata commands for the estimation of parametric stochastic frontier (SF) models using cross-sectional data (Bell and Bellotti 2014; Newton et al. 2014). Table 3 shows the maximum likelihood estimates for parameters of the stochastic frontier model. Almost all inputs have positive correlation with maize profitability except for fertilizer, herbicide, and labor costs that have negative effects on maize output variable.

Table 4 shows the determinants of technical inefficiency in maize production. Inefficiency is the dependent variable in the technical inefficiency model, and as such, variables with a negative (positive) coefficient sign will have a positive (negative) impact on technical efficiency. The analysis found that frequency of extension contact had a negative and significant effect on inefficiency. This implies that farmers with high frequency of extension contact are more technically efficient. Extension officers impart skills to farmers through one-on-one visits, training workshops, advisory services, and promotional events like exchange visits and field days. Farmers can thus learn about new technologies when they are in constant contact with extension, and thus they end up becoming more efficient farmers. This finding is in line with those of Dziwornu and Sarpong (2014), Welch et al. (2016), and Abdulai et al. (2018).

They are also in line with findings from Mango et al. (2015) who found a negative and statistically significant relationship between technical efficiency and extension contact in smallholder farming systems of Zimbabwe following the fast track land reform program. Another researcher, Konja et al. (2019), also found positive impact of extension contact on technical efficiency in certified groundnut seed production in Northern Ghana.

Correspondingly, the coefficient for number of locally available traders was negative and significant. This means that farmers who have access to farm gate traders are technically efficient. Maize farmers in most rural areas are constrained when it comes to capital and hence have difficulties to access distant markets. Therefore, if traders come to buy locally, this acts as an incentive for them to produce that crop knowing there is going to be a guaranteed market with potential to lower transaction costs. Furthermore, the coefficient of CSA adoption was negative and significant. This means that farmers using CSA technologies are more efficient.

The stochastic frontier results showed that fertilizer and other costs have negative and significant effect on the inefficiency of maize profitability. The negative signs of the variables indicate that as these variables increase, the profit inefficiency of maize producers decreases. This means a unit increase in costs of the basal dressing fertilizer (DFERT) and top dressing (ANFERT) will lead to 166.68% and 40.02% increases in profitability, respectively. Basal and top dressing fertilizer applications are very critical for maize profitability, and the increase in use as proxied by costs will result in increased profitability. Total livestock units (TLU) and farming experience had significant positive coefficients implying that as the farmer's TLU/head size and farming experience increase, the profit inefficiency of the farmers also increases. This contradicts prior expectation and might be explained by the fact that experienced farmers are older and unwilling to invest in any new technologies that come around.

Conclusions and Recommendations

The most economic adaptation strategy in the face of climate change would be adoption of CSA technologies as evidenced by positive gross margins and higher returns on investment when compared to the conventional way of farming. This is further supported by the positive effect of CSA adoption on technical efficiency. Farmers should however note that not all adaptation strategies are economical; thus, record-keeping of costs and income for regular computation of costs and benefits is crucial. Farmers can then choose technologies that give higher benefits or those that use less inputs given that most of the farmers are financially constrained. Based on variables that significantly influenced profit efficiency, the study makes three recommendations.

Government should continue putting resources towards supporting mobility of extension staff for continued extension to farmer contact and giving them adequate resource (information materials) so that they continue delivering key information on yield enhancing CSA technologies.

Policies to promote inorganic fertilizer use in order to boost soil fertility remain critical. Government should therefore strengthen the capacity of rural agro dealers to sell fertilizers locally at reasonable prices.

Policies to promote farm gate buying or market centers within wards should also be put in place as they have the potential to increase efficiency if farmers are aware of such a guaranteed market with very low transaction costs.

Declaration This research did not receive any specific grant from funding agencies in the public, commercial, or not-for-profit sectors.

Declaration of Competing Interest The authors certify that they have no affiliations with or involvement in any organization or entity with any financial interest (such as honoraria; educational grants; participation in speakers' bureaus; membership, employment, consultancies, stock ownership, or other equity interest; and expert testimony or patent-licensing arrangements) or non-financial interest (such as personal or professional relationships, affiliations, knowledge, or beliefs) in the subject matter or materials discussed in this manuscript.

Acknowledgment The authors wish to appreciate the Ministry of Agriculture officials and Rural District Administrators for granting the permission to do data collection in the sampled districts.

References

Abate, Tsedeke, Monica F, Tahirou A, Girma TK, Rodney L (2017) Characteristics of Maize Cultivars in Africa : How Modern Are They and How Many Do Smallholder Farmers Grow? Agriculture & Food Security 6(30):1–17. https://doi.org/10.1186/s40066-017-0108-6

Abdulai S, Nkegbe PK, Donkoh SA, Yildiz F (2018) Assessing the technical efficiency of maize production in northern Ghana: the data envelopment analysis approach. Cogent Food Agric 4 (1):1–14. https://doi.org/10.1080/23311932.2018.1512390

Ali A, Erenstein O (2017) Climate risk management assessing farmer use of climate change adaptation practices and impacts on food security and poverty in Pakistan. Clim Risk Manag 16:183–194. https://doi.org/10.1016/j.crm.2016.12.001

Bell LW, Bellotti B (2014) Whole-farm economic, risk and resource-use trade-offs associated with integrating forages into crop – livestock systems in . . .. Agric Syst 133:63–72. https://doi.org/10.1016/j.agsy.2014.10.008

Dziwornu RK, Sarpong DB (2014) Application of the stochastic profit frontier model to estimate economic efficiency in small-scale broiler production in the Greater Accra region of Ghana. Rev Agric Appl Econ 17(02):10–16. https://doi.org/10.15414/raae.2014.17.02.10-16

FAOSTAT (2020) Food and Agriculture Data. Food and Agriculture Organization of the United Nations. Statistics Division, Rome. http://fenix.fao.org/faostat/internal/en/#data/QC

GOZ (2018) National Agriculture Policy Framework 2019–2030. Ministry of Lands, Agriculture, Water, Climate and Rural Resettlement, Harare

Greene W (2010) A stochastic frontier model with correction for sample selection. J Prod Anal 34 (1):15–24. https://doi.org/10.1007/s11123-009-0159-1

Kassie GT, Abdulai A, Greene WH, Shiferaw B, Abate T, Tarekegne A, Sutcliffe C (2017) Modeling preference and willingness to pay for drought tolerance (DT) in maize in rural Zimbabwe. World Dev 94:465–477. https://doi.org/10.1016/j.worlddev.2017.02.008

Konja T, Dominic FNM, Oteng-Frimpong R (2019) Profitability and profit efficiency of certified groundnut seed and conventional groundnut production in northern Ghana: a comparative analysis. Cogent Econ Financ 7(1). https://doi.org/10.1080/23322039.2019.1631525

Mango N, Makate C, Hanyani-Mlambo B, Siziba S, Lundy M (2015) A stochastic frontier analysis of technical efficiency in smallholder maize production in Zimbabwe: the post-fast-track land reform outlook. Cogent Econ Financ 3(1):1–14. https://doi.org/10.1080/23322039.2015.1117189

Michler JD, Baylis K, Arends-Kuenning M, Mazvimavi K (2019) Conservation agriculture and climate resilience. J Environ Econ Manag 93:148–169. https://doi.org/10.1016/j.jeem.2018.11.008

Mujeyi A (2018) Policy and institutional dimensions in climate-smart agriculture adoption: case of rural communities in Zimbabwe. In: Filho WL (ed) Handbook of climate change resilience. Springer Nature Switzerland AG. https://doi.org/10.1007/978-3-319-71025-9_66-1

Mujeyi K, Mujeyi A (2018) Fostering climate smartness in smallholder farming systems: business promotional approaches for improved maize varieties in eastern and southern Africa. In: Leal Filho W (ed) Handbook of climate change resilience. Springer International Publishing AG, part of Springer Nature. https://doi.org/10.1007/978-3-319-71025-9_24-1

Karakaplan MU (2017) The Stata Journal. In: The Stata, edited by Joseph H Newton and Nicholas J Cox. Vol. 17. College Station, Texas, USA: The Stata Press

Newton HJ, Cox NJ, Nichols A, Dc W, Gilmore L (2014) The Stata Journal. Stata J 4:778–797

Nsikak-Abasi AE, Okon S (2013) Sources of technical efficiency among smallholder maize farmers in Osun state of Nigeria. Discourse J Agric Food Sci 1(4):48–53. http://docsdrive.com/pdfs/medwelljournals/rjasci/2010/115-122.pdf

Omhile T, Renato V, David H (2016) Evaluating the productivity gap between commercial and traditional beef production systems in Botswana. AGSY 149:30–39. https://doi.org/10.1016/j.agsy.2016.07.014

Papendiek F, Tartiu VE, Morone P, Venus J (2016) Assessing the economic profitability of fodder legume production for green biorefineries – a cost-benefit analysis to evaluate farmers profitability 112:3643–3656. https://doi.org/10.1016/j.jclepro.2015.07.108

Poole N (2017) Smallholder agriculture market participation. https://doi.org/10.3362/9781780449401

Rosenstock TS, Rohrbach D, Nowak A, Girvetz E, Tracking Progress. (2019) The Climate-Smart Agriculture Papers. The Climate-Smart Agriculture Papers. https://doi.org/10.1007/978-3-319-92798-5

Rurinda J, Mapfumo P, Van Wijk MT, Mtambanengwe F, Rufino MC (2014) Climate risk management sources of vulnerability to a variable and changing climate among smallholder households in Zimbabwe: a participatory analysis. Clim Risk Manag 3:65–78. https://doi.org/10.1016/j.crm.2014.05.004

Sain G, María A, Corner-dolloff C, Lizarazo M, Nowak A, Martínez-barón D, Andrieu N (2017) Costs and benefits of climate-smart agriculture: the case of the dry corridor in Guatemala. Agric Syst 151:163–173. https://doi.org/10.1016/j.agsy.2016.05.004

Shongwe P, Masuku MB, Manyatsi AM (2014) Cost benefit analysis of climate change adaptation strategies on crop production systems: a case of Mpolonjeni Area Development Programme (ADP) in Swaziland 3(1):37–49. https://doi.org/10.5539/sar.v3n1p37

Siyuan S, Tong L, Liu RH (2018) Corn phytochemicals and their health benefits. Food Sci Human Wellness 7(3):185–195. https://doi.org/10.1016/j.fshw.2018.09.003

Steward PR, Dougill AJ, Thierfelder C, Pittelkow CM, Stringer LC, Kudzala M, Shackelford GE (2018) The adaptive capacity of maize-based conservation agriculture systems to climate stress in tropical and subtropical environments: a meta-regression of yields. Agric Ecosyst Environ 251(2017):194–202. https://doi.org/10.1016/j.agee.2017.09.019.

Thierfelder C, Chivenge P, Mupangwa W, Rosenstock TS, Lamanna C, Eyre JX (2017) How climate-smart is conservation agriculture (CA)? – its potential to deliver on adaptation, mitigation and productivity on smallholder farms in Southern Africa. https://doi.org/10.1007/s12571-017-0665-3

UNCCS (2019) Climate action and support trends. United Nations Climate Change Secretariat, Bonn

Villano R, Fleming E (2006) Technical inefficiency and production risk in Rice farming: evidence from Central Luzon Philippines. Asian Econ J 20(1):29–46. https://doi.org/10.1111/j.1467-8381.2006.00223.x

Waddington SR, Karigwindi J, Chifamba J (2007) The sustainability of a groundnut plus maize rotation over 12 years on smallholder farms in the sub-humid zone of Zimbabwe. African Journal of Agricultural Research. 2(8):342–348

Wang H-J (2008) Stochastic frontier models. In: The new Palgrave dictionary of economics, pp 925–928. https://doi.org/10.1057/9780230226203.1623

Welch EW, Villanueva AB, Jha Y, Ogwal-omara R, Welch E, Wedajoo S, Halewood M (2016) Adoption of climate smart technologies in East Africa findings from two surveys and participatory exercises with farmers and local and experts. https://doi.org/10.13140/RG.2.2.24562.09927

9

Pyrolysis Bio-oil and Bio-char Production from Firewood Tree Species for Energy and Carbon Storage in Rural Wooden Houses of Southern Ethiopia

Miftah F. Kedir

Contents

Introduction
Material and Methods
    Description of the Study Area
    Methods of Data Collection and Analyses
Results
    Bio-oil and Bio-char Yield of Different Woody Biomasses Species
    Calorific Value and Moisture Content of Bio-oil and Bio-char in Comparison
    with the Parent Firewood 1
    Fixed Carbon, Volatile Matter, and Moisture Content of Different Tree Species Bio-
    chars
    The Potential of Pyrolysis Oil and Bio-char in Reducing Firewood End Use
    Emission
Discussion
Conclusion and Recommendation
References

Abstract

The need for emission reduction for climate management had triggered the application of pyrolysis technology on firewood that yield bio-oil, bio-char, and syngas. The purpose of present study was to select the best bio-oil and bio-char producing plants from 17 firewood tree species and to quantify the amount of carbon storage. A dried and 1 mm sieved sample of 150 g biomass of each species

M. F. Kedir (✉)
WGCFNR, Hawassa University, Shashemene, Ethiopia

Central Ethiopia Environment and Forest Research Center, Addis Ababa, Ethiopia was pyrolyzed in assembled setup of tubular furnace using standard laboratory techniques. The bio-oil and bio-char yields were 21.1–42.87% (w/w) and 23.23–36.40% (w/w), respectively. The bio-oil yield of *Acacia seyal*, *Dodonea angustifolia*, *Euclea schimperi*, *Eucalyptus globulus*, *Casuarina equisetifolia*, and *Grevillea robusta* were over 36% (w/w), which make the total yield of bio-oil and bio-char over 62% (w/w) of the biomass samples instead of the 12% conversion efficiency in traditional carbonization. The calorific value of firewood was 16.31–19.66 MJ kg^{-1} and bio-oil was 23.3–33.37 MJ kg^{-1}. The use of bio-oil for household energy and bio-char for carbon storage reduced end use emission by 71.48–118.06%, which could increase adaptation to climate change in comparison to open stove firewood by using clean fuel and reducing indoor pollution.

Keywords

Bio-char · Bio-oil · Deforestation · End use emission · Woody biomass

Introduction

The adaptation and mitigation of climate change and energy security requires alternative energy sources that reduce emission of greenhouse gasses (GHG) in the place of fossil fuel dominant economy of the world (Krajnc et al. 2014). Liquid biofuel production from biomass pyrolysis is new form of the old technology that reduces waste and improves the low bulk density, high moisture content, hydrophilic nature, and low calorific value of firewood (Arias et al. 2008). Pyrolysis is a thermochemical process of converting biomass in to solid bio-char, liquid bio-oil (also called pyrolysis oil), and syngas in the absence of oxygen at 300–1000 °C, heating rate 0.1–1000 °C s^{-1}, and vapor residence time 0.5–1800 s (Granada et al. 2013). Pyrolysis of dry biomass ($C_6H_{10}O_5$) produces combustible gases (H_2, CH_4, CO) and noncombustible gases (CO_2 and H_2O) resulting in condensable gases forming bio-oil (C_6H_8O with H_2O) (Cordeiro 2011). High heating rates above 500 °C and short vapor residence time gives more pyrolysis oil; and low temperature below 400 °C produces more bio-char (Xiu and Shahbazi 2012).

Feedstocks for pyrolysis can be a variety of woody and non-woody biomasses, forest products, solid organic wastes, forest/agricultural residues, paper and cardboard except toxic biomasses that have heavy metals, polyaromatic hydrocarbons, and dioxins (Garcia-Perez 2008).

The calorific value of bio-oil can be chemically upgraded to 44 MJ kg^{-1} (Elliot 2012). Firewood combustion pollutes indoor air and affects health but bio-oil has no significant health, environment, or safety risks and its GHG emission is lower than petro diesel and gasoline (Shimelis 2011).

Bio-char and charcoal are similar products of pyrolysis technology that are used for different purpose. Bio-char is charcoal like, fine porous structured, positively charged, high carbon and low moisture containing co-product of bio-oil but charcoal is coarse structure as a sole product. Bio-char could be used for energy supply, soil carbon storage, and fertility amendment but charcoal is usually used for energy.

The pyrolysis of organic matter alters the chemical structure of carbon to aromatic carbon rings called recalcitrant carbon that resist microbial decomposition. Wood bio-char stores 25–50% of its carbon for millennia, 100–1000 years but organic residue compost stores 10–20% of its carbon from weeks to 5–10 years (Kannan et al. 2013).

Capturing the volatiles during pyrolysis to get bio-oil, in addition to bio-char and syngas instead of mere charcoal, is an increment of conversion efficiency of carbonization in traditional charcoal making (Brown 2009). Charcoal making emits primary GHG of the energy system, carbon monoxide, ethane, pyroacids, tars, heavy oils, and water (Bird et al. 2011). Firewood and charcoal have CO_2 emission factors of 112,000 kg TJ^{-1} on net calorific value basis (IPCC 2006). About 1788±337 g CO_2 and 32±5 g CH_4 per kilogram of charcoal are produced, which varies with different vegetation parts and burning conditions (Chidumayo and Gumbo 2013). In charcoal making the three steps include wood sourcing, carbonization, and end use, and the emission is 29–61%, 28–61%, and 9–18%, respectively (FAO 2017).

Climate change adaptation to energy is the adjustment in natural or human systems in response to actual or expected energy deficit of climatic stimuli or their effects, which moderates harm or exploits beneficial energy production opportunities. Ethiopia in particular and Africa in general have low adaptation capability to climate change (IPCC 2006). The conversion of firewood and charcoal in to multiple products of bio-char, bio-oil, and syngas reduces consumption of biomass that reduces deforestation and increases income sources in order to adapt to climate change. In Ethiopia and elsewhere in Africa, adapting technology for alternative sources of bioenergy is one of the strategies of climate change adaptation (FAO 2017). Environmental sustainability could be achieved by local management of firewood saving and macro policy adjustment in order to promote the sustainability of land resources and climate change adaptation (Eze et al. 2020). Therefore, developing biomass saving technologies like pyrolysis is important for climate change adaptation and mitigation.

Pyrolysis oil during biomass carbonization can be produced over a wider range of temperature above 300 °C by screening a large number of biomass yielding trees (Xiu and Shahbazi 2012). According to Okoroigwe et al. (2015) tropical woody biomass produces up to 66% (w/w) bio-oil for energy and it contains industrially useful chemicals. Bio-oil production is an attractive venture with significant commercial application and value, and dry feed can produce up to 80% (w/w) bio-oil. However, there is dearth of information on condensing the volatile matters to bio-oil, especially at 600 °C and residence time of 2 s in Ethiopia for firewood species, except *Catha edulis* (Yishak 2014). In fact, it is studied that biomass residues at 600 ° C has high recalcitrant character and low volatile nature (Jindo et al. 2014). One of the most important characteristics of biomass fuel is heating value which can be determined experimentally by adiabatic bomb calorimeter (Sheng and Azevedo 2005), which was not available for many of the firewood tree species in Ethiopia. The purpose of the present study was to inform tree selection in plantation development by the bio-oil and bio-char yield, and by their carbon storage potential in selected firewood utilizing rural households of Southern Ethiopia.

Material and Methods

Description of the Study Area

Biomass samples of fire wood species were collected in Southern Ethiopia, three agro-ecologies of Enemorina Ener district (county). From each agro-ecology a representative peasant association (PA) or Kebele (lowest administrative unit) was sampled. In lowland agro-ecology, 500–1600 m altitude above sea level (asl), Ener Kola PA; in midaltitude agro-ecology, 1600–2400 m altitude asl, Daemir PA; and in highland agro-ecology, 2400–3200 m asl, Awed PA were selected (Fig. 1). The local

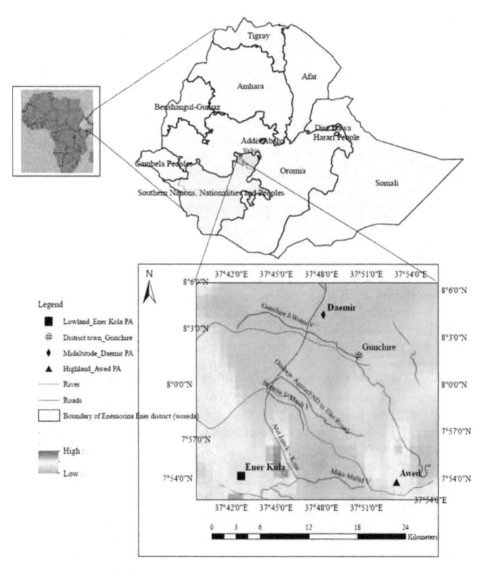

Fig. 1 Location map of most samples collected areas. (The legend and scale refers only to the three studied peasant associations or Kebeles and nearby features)

people had been practicing mixed farming, trading, and pottery work. The main fuel for cooking was firewood with or without kerosene light.

Methods of Data Collection and Analyses

Field Sample Collection

Woody biomass samples were collected from the selected PAs after interviewing 5–10 key informants, and district energy offices about the preferred firewood tree species that ranked 1–5. The most preferred tree species were selected for pyrolysis test in each agro-ecology; seven species in lowland and midaltitude each and one species in highland. Additional samples from two species (*Catha edulis* and *Prosopis juliflora*) were obtained from other places (Table 1) because of their abundance and common firewood value for comparison purposes, making a total of 17 species. That is *C. edulis* is abundant throughout Ethiopia and *P. juliflora* is an invasive species used for charcoal making in North Eastern Ethiopia. The family, diameter at breast height (DBH), and total mean height (*H*) of the selected species are given in Table 1.

From the 15 tree species in the forest, three standing trees were randomly selected and cut from each. Then the wood without debarking was chipped in to 1–5 cm long pieces by excluding branches (≤5 cm diameter) and leaves. In *C. edulis*, leafy residues and in *C. megalocarpus,* fruit pod were collected. Then the 17 tree species biomass samples were air dried, separated from any impurities, ground, and sieved by 1 mm sieve size.

Experimental Analyses

Pure, composite, ground, dry biomass samples of 150 g were pyrolyzed in assembled setup of tubular furnace in three replications. The pyrolysis was done at 600 °C temperature, at a heating rate of 100 °C min^{-1} and 2 s vapor residence, using 1.5 atmosphere inert nitrogen gas with 20–25 mL min^{-1} flow rate. The setup consisted of feeder, reactor, glass liquid collecting condenser, and chiller (Fig. 2). Only bio-oil and bio-char were collected. The weight of bio-char and bio-oil was measured with balance (Adam Lab.equipment Leicester LE67FT-England, 0.001 g) and volume by graduated cylinder. The yield of bio-char and bio-oil was determined from the proportion of dried biomass feedstock pyrolyzed (Eq. 1) and loss by deduction (Eq. 2). The percentage throughout this chapter is given as (%) for percent weight to weight (%, w/w) or % (w/w) or %, unless otherwise specified as percent volume to volume as (%, v/v) or % (v/v).

$$\textbf{Yield of biooil or biochar}\left(\%, \frac{\mathbf{w}}{\mathbf{w}}\right) = \left(\frac{\textbf{\textit{W} of biooil or biochar (g)}}{\textbf{\textit{W} of pyrolysed feedstock (g)}}\right)$$
$$\times \, \mathbf{100} \qquad (1)$$

$$\textbf{Biochar}\left(\%, \frac{\mathbf{w}}{\mathbf{w}}\right) + \textbf{Biooil}\left(\%, \frac{\mathbf{w}}{\mathbf{w}}\right) + \textbf{NCG}\left(\%, \frac{\mathbf{w}}{\mathbf{w}}\right) = \mathbf{100}\left(\%, \frac{\mathbf{w}}{\mathbf{w}}\right) \qquad (2)$$

Table 1 Description of sampled tree species for firewood, bio-oil, and bio-char production

| Species | Family | Sample of collection | | | Sources |
		Agro-ecology	Mean DBH (cm)	Mean H (m)	
Acacia albida (Delile) Chev.	Fabaceae	Lowland	16.07	12.3	Bekele (2007)
Acacia seyal Delile	–	Lowland	17.8	14.1	–
Acokanthera schimperi (A. DC.) Schweinf.	Apocynaceae	Lowland	10.73	7.9	–
Combretum collinum, Fresen.	Combretaceae	Lowland	10.08	6.9	–
Euclea schimperi (A.DC.) Dandy	Ebenaceae	Lowland	11.97	10.5	–
Casuarina equisetifolia L.	Casuarinaceae	Lowland	21.3	15.5	–
Dodonaea angustifolia L.f.	Sapindaceae	Lowland	4.12	4.00	–
Acacia abyssinica Hochst. ex Benth	Fabaceae	Midaltitude	24.5	12.0	–
Acacia decurrens Willd.	Fabaceae	Midaltitude	27.5	11.5	–
Cupressus lusitanica Mill.	Cupressaceae	Midaltitude	24.6	15.5	–
Catha edulis (Vahl.) Endl.	Celastraceae	Midaltitude	–	–	–
Eucalyptus camaldulensis Dehnh.	Myrtaceae	Midaltitude	21.5	16.0	–
Grevillea robusta A.Cunn. ex R.Br.	Proteaceae	Midaltitude	20.5	14.5	–
Pinus patula Schldl. et Cham.	Pinaceae	Midaltitude	19.3	15	–
Eucalyptus globulus Labill.	Myrtaceae	Highland	27.3	12.5	–
Prosopis julifolra (Sw.) DC.	Fabaceae	Afar	22.5	17.5	–
Croton megalocarpus Hutch.	Euphorbiaceae	Hawassa city	10.5	6.5	Aliyu et al. (2010)

where W is weight; NCG is non-condensable gas by considering that all condensable gases were condensed.

The bio-oil was degummed in 3% (v/v) distilled water and centrifuged at 2000 rpm for 20 min. The moisture content was determined gravimetrically by taking 5 g bio-oil by heating at 105 °C in oven up to constant weight (Eq. 3).

$$\mathbf{MC}\left(\%, \frac{\mathbf{w}}{\mathbf{w}}\right) = \left(\frac{(W \text{ of initial sample } (g) - W \text{ of sample at } 105°\mathbf{C}(g))}{W \text{ of initial sample } (g)}\right) \times \mathbf{100}$$

(3)

where MC is moisture content; W is weight.

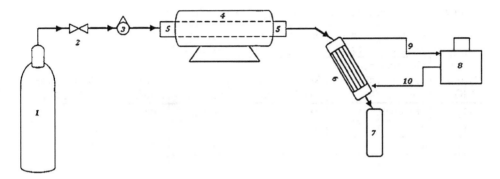

Fig. 2 Simplified system of pyrolysis, assembled setup of tubular furnace
Legend (not to scale):
1. N_2 gas containing cylinder; 2. Valve; 3. Pressure gauge; 4. Tubular furnace; 5. Stainless steel tube (inserted into the furnace); 6. Glass condenser; 7. Condensate collector; 8. Chiller; 9. Hot water to chiller; 10. Cooling water (at 4 °C) to condenser

The bio-oil and parent firewood Gross Calorific Value (GCV) was determined by adiabatic oxygen bomb calorimeter (Parr® model no.1241EF adiabatic calorimeter S.no. 5172, 115 v, 50 Hz, 2.0 Amps, Parr Instrument Company). About 1 g sample and oxygen filled bomb at 30 atmospheres were used. GCV was converted to net calorific value (NCV) by multiplying with 0.95 and 0.80 for oil and dry woody biomass, respectively (Forest Products Laboratory 2004). Ash content of firewood and bio-oil was determined by burning 5 g sample in furnace at 600 °C for 4 h.

Bio-char samples were characterized using proximate analysis (ASTM 1989) to obtain fixed carbon, moisture content (D3173), ash content (D3174), and volatile matter (D3175). Moisture content of bio-char was determined by drying 1 g initial sample in oven at 105 °C to constant weight for 3 h (Eq. 3); volatile matter by heating 1 g moisture free sample at 950 °C for 6 min (Eq. 4); ash content by heating 1 g sample at 750 °C for 3 h (Eq. 5) and fixed carbon by deduction (Eq. 6).

$$\mathbf{Vm}\left(\%, \frac{\mathbf{w}}{\mathbf{w}}\right) = \left(\frac{\textit{W of sample at } 105°C \ (g) - \textit{W of sample at } 950°C(g)}{\textit{W of initial sample } (g)}\right) \times \mathbf{100}$$

$$(4)$$

$$\mathbf{Ash} \ (A)\left(\%, \frac{\mathbf{w}}{\mathbf{w}}\right) = \left(\frac{\textit{W of sample at } 750°C \ (g)}{\textit{W of initial sample } (g)}\right) \times \mathbf{100} \qquad (5)$$

$$\mathbf{Fixed\ carbon\ of\ biochar}\left(\%, \frac{\mathbf{w}}{\mathbf{w}}\right) = (\mathbf{100} - (\mathbf{MC}, \% + \mathbf{Vm}, \% + A, \%)) \qquad (6)$$

where MC is moisture content; W is weight; Vm is volatile matter; A is ash content

The common GHG (CO_2, CH_4, and N_2O) emissions in rural wooden houses were determined based upon NCV (IPCC 2006) of bio-oil and firewood with or without kerosene. Although emission exists throughout the life cycles of these fuels, from

Table 2 Emission factors and global warming potential of common GHGs from different fuels

| Type of GHG | Residential source of Gas Emission Factors on a net calorific basis (Stationary Combustion) (IPCC 2006) (kg TJ^{-1}) | | | | GWP (100-year time horizon) |
	Firewood in conventional stove	Bio-oil (other liquid biofuels)	Charcoal	Other kerosene	
CO_2	112000	79600	112000	71900	1
CH_4	210	10	330.5	10	23
N_2O	4	0.6	1.6	0.6	296

planting, woody biomass harvesting to consumption, the present study's concern of indoor pollution reduction dealt the emission only at the end use or consumption. The IPCC default emission factors were applied (Eqs. 7 and 8). The GHGs were converted to carbon dioxide equivalent (CO_2^{-e}) using the global warming potential (GWP) of each gas (Table 2).

The average end use emission of firewood and bio-oil combustion and bio-char soil carbon accumulation was determined. For comparison purpose, the daily consumption of firewood and kerosene at household level in each agro-ecology was obtained from Miftah et al. (2017). It was assumed that a household consumes the same amount of energy in using firewood alone, firewood with kerosene or bio-oil.

The carbon storage of bio-char was determined by its fixed carbon content (Eq. 9). Stable carbon fraction to be stored for over a century used 80% factor (Roberts et al. 2010). In the presence of modern pyrolysis reactor 75% bio-oil and 12% bio-char yield was assumed to be produced from woody biomass (Granada et al. 2013). The total energy of the fuels was determined by the dried weight of the fuel and their calorific value (ASTM 1989).

The emission from firewood combustion in conventional wood stove using IPCC (2006) was calculated as Eq. 7.

$$\textbf{Wd em} = \sum_{i}^{k} \textbf{Wd cons}_{(ijk)} \times \textbf{EF}_{(ijk)} \times \textbf{GWP}_{(ijk)} \qquad (7)$$

where Wd em is emission of GHG ($kgCO_2^{e-}$) from firewood combustion; Wd cons is firewood consumption in net calorific value ($TJkg^{-1}$); EF is emission factor of firewood on net calorific value basis ($kgTJ^{-1}$); and GWP is global warming potential of a given gas on 100 years; i is CO_2; j is CH_4, and k is N_2O.

The emission from bio-oil combustion in stove using IPCC (2006) was calculated as Eq. 8.

$$\textbf{Bo em.} = \sum_{i}^{k} \textbf{Bo cons}_{(ijk)} \times \textbf{EF}_{(ijk)} \times \textbf{GWP}_{(ijk)} \qquad (8)$$

where Bo em is emission of GHG ($kgCO_2^{e-}$) from bio-oil combustion; Bo cons is bio-oil consumption in net calorific value ($TJkg^{-1}$); EF is emission factor of bio-oil on net calorific value basis ($kgTJ^{-1}$); and GWP is global warming potential of a given gas on 100 years basis; i is CO_2; j is CH_4 and k is N_2O.

Carbon storage in bio-char of organic wastes was calculated as Eq. 9.

$$\textbf{Csb } (\textbf{g}) = (\textbf{dry biomass } (\textbf{g})) \times (\textbf{biochar yield } (\%)) \times (\textbf{fixed } \textbf{\textit{C}} (\%)) \qquad (9)$$

where Csb is carbon storage in bio-char.

Results

Bio-oil and Bio-char Yield of Different Woody Biomasses Species

The bio-oil (Fig. 3) yield of woody biomass samples ranged from 23.03 (%, w/w) in *A. schimperi* to 42.9 (%, w/w) in *E. globulus* which was statistically different at $p<0.05$ (Table 3). The bio-oil yield of fruit pod of *C. megalocarpus* was about 21.1 (%, w/w), lower than the other woody biomasses; and the leaf of *C. edulis* was intermediate about 25.83 (%, w/w). Tree species like *E. globulus, A. seyal, D. angustifolia, E. schimperi*, and *G. robusta* produced greater amount of bio-oil (Table 3) and highly preferred for firewood (Table 3).

The bio-char (Fig. 4) yield of the biomass samples ranged from 23.2% (w/w) in *E. camaldulensis* to 36.4% (w/w) in *C. edulis* which was statistically different at $p< 0.05$ (Table 3). The mass losses of pyrolysis product ranged from 31.23% (w/w) in *E. schimperi* to 46.37% (w/w) in *A. schimperi* (Table 3), which could be attributed to the specific characteristics of the species.

Calorific Value and Moisture Content of Bio-oil and Bio-char in Comparison with the Parent Firewood

The calorific value of firewood used for the pyrolysis process ranged from 16.31 $MJkg^{-1}$ (in *E. schimperi*) to 19.66 $MJkg^{-1}$ (in *P. julifolra*) at moisture content

Fig. 3 Bio-oil from pyrolysis of woody biomass

Table 3 Bio-oil and bio-char yield of different woody plant species

Species	Yield (Mean±Stand. err.) (%, w/w)		Priority as firewood[g]
	Bio-oil	Bio-char	
A. abyssinica	29.067±3.254[abcd]	31.367±1.717[bc]	1
A. albida	33.433±1.802[abcd]	30.567±0.769[bc]	1
A. decurrens	33.533±3.641[abcd]	30.233±0.555[bc]	2
A. schimperi	23.033±2.567[ab]	30.600±0.794[bc]	4
A. seyal	39.000±1.808[cd]	27.133±0.736[ab]	1
C. collinum	31.633±3.537[abcd]	30.933±1.212[bc]	1
C. edulis[e]	25.833±2.634[abc]	36.400±0.208[d]	5
C. equisetifolia	36.067±4.390[bcd]	27.100±0.173[ab]	1
C. lusitanica	34.167±2.195[abcd]	27.500±0.173[ab]	3
C. megalocarpus[f]	21.100±1.200[a]	33.300±1.700[c]	3
D. angustifolia	38.033±0.219[cd]	24.633±0.186[a]	1
E. camaldulensis	32.767±1.139[abcd]	23.233±0.841[a]	1
E. globulus	42.867±0.888[d]	25.733±1.033[a]	1
E. schimperi	37.133±2.118[cd]	31.633±1.650[bc]	1
G. robusta	38.800±3.623[cd]	25.267±0.219[a]	2
P. juliflora	31.367±4.296[abcd]	30.767±0.536[bc]	3
P. patula	27.800±1.600[abc]	27.600±1.900[ab]	3

Note: [a, b, c,] and [d] are statistically different at $p<0.05$ of ANOVA
[e]Leaves
[f]Fruit pod
[g]Priority 1 is highly preferred and 5 is not

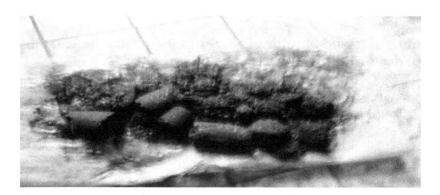

Fig. 4 Bio-char from pyrolysis of woody biomass

of 7.92–10.22 (%, w/w) (Fig. 5). Firewood from *D. angustifolia, E. camaldulensis, G. robusta, A. decurrens*, and *E. globulus* had calorific value above 18 MJ kg^{-1}. *C. edulis* leafy residues had comparable calorific value with the other firewood species (Fig. 5), but not used as firewood (Table 3) because of its smoke. The calorific value of bio-oil ranged from 21.43 MJkg^{-1} (in *A. albida*) to 33.37 MJkg^{-1} (in *E. globulus*) (Fig. 5).

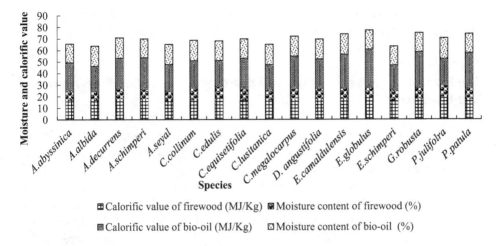

Fig. 5 The mean calorific value of parent firewood, bio-char, and bio-oil of a given plant species

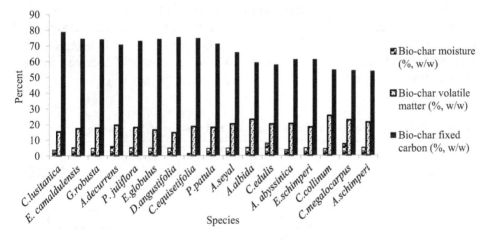

Fig. 6 Mean fixed carbon, volatile, and moisture content of bio-char

Fixed Carbon, Volatile Matter, and Moisture Content of Different Tree Species Bio-chars

The fixed carbon content of bio-chars of the different tree species ranged from 53.48 (%, w/w) (in *A. schimperi*) to 78.85 (%, w/w) (in *C. lusitanica*); and volatile matter from 14.52 (%, w/w) (in *D. angustifolia*) to 25.31 (%, w/w) (*C. collinum*) (Fig. 6).

Ash Content of Firewood Feedstock, and Bio-oil and Bio-char

The percentage of the ash content of the different tree species firewood and bio-oil was not consistently increasing or decreasing (Table 4). The ash in bio-char ranged from 2.27(%, w/w) in *C. lusitanica* to 20.65 (%, w/w) in *A. schimperi*. In firewood the ash content ranged from 0.434 (%, w/w) in *E. camaldulensis* to 8.418 (%, w/w) in *E. schimperi*. The lowest proportion of ash was obtained in bio-oil, ranging from

0.135% in *G. robusta* to 0.892% in *C. equisetifolia*. As the ash content of the firewood increases the ash content of the bio-char also increased (Table 4).

The Potential of Pyrolysis Oil and Bio-char in Reducing Firewood End Use Emission

Rural households were using firewood and kerosene as energy sources. In the presence of kerosene, the amount of firewood biomass used was reduced, but both biomass and kerosene together during consumption emitted annually about 2.323–4.509 t CO_2^e in each household. The end use emission from fire wood combustion was 2.335–4.527 t CO_2^e year^{-1} (Fig. 7), which could be reduced to 1.88–3.645 t CO_2^e year^{-1} in bio-oil heating in each household of the studied PAs. Moreover, bio-oil production has corresponding bio-char that can store carbon 1.214–4.363 t CO_2^e year^{-1} (Fig. 7) that makes the net emission of 0.67 to net storage of 0.72 t CO_2^e year^{-1}. The pyrolysis in the present study that produced bio-oil yield of 21–42.87(%, w/w) as alternative energy for a household cooking and bio-char for carbon storage reduced end use emission by 71.48–118.06% in each household of the PAs when compared with the emission by firewood (Fig. 7). In using bio-char that produced 23.23–36.4%, w/w (Table 3) from a household firewood for soil amelioration can store 0.27–0.95 t C year^{-1} for over a century (Fig. 8).

Table 4 The mean ash content of firewood, bio-oil, and bio-char of different tree species

Species	Mean ash (%, w/w)		
	Firewood	Bio-oil	Bio-char
A. abyssinica	4.586	0.531	15.03
A. albida	4.421	0.567	12.80
A. decurrens	1.418	0.688	3.72
A. schimperi	4.382	0.758	20.65
A. seyal	3.471	0.794	9.45
C. collinum	7.358	0.588	16.11
C. edulis[a]	7.816	0.787	14.51
C. equisetifolia	1.788	0.892	5.36
C. lusitanica	1.049	0.685	2.27
C. megalocarpus[b]	6.159	0.724	16.38
D. angustifolia	1.506	0.727	5.17
E. camaldulensis	0.434	0.618	3.15
E. globulus	3.594	0.352	4.54
E. schimperi	8.418	0.775	15.98
G. robusta	0.742	0.135	3.50
P. juliflora	1.748	0.831	3.87
P. patula	7.459	0.146	6.36

[a]Leaves
[b]Fruit pod

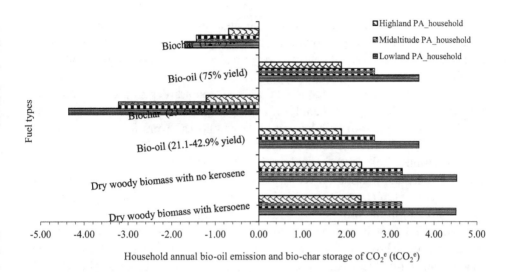

Fig. 7 Emission and carbon storage of fuels at household level with similar gross heating value. (Note: Negative x-axis indicates the CO_2^e stored in bio-char; positive X-axis indicates the CO_2^e emission)

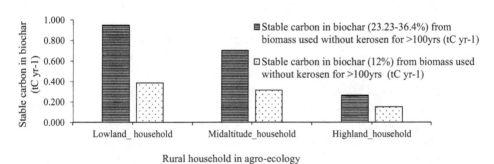

Fig. 8 Stable carbon storage in bio-char in converting firewood to bio-oil

Discussion

The use of pyrolysis technology to reduce the wastage of biomass and diversify products like biochar, bio-oil, and syngas increases adaptation and mitigation to climate change (Shackley et al. 2012). Greater bio-oil yield was obtained from locally preferred firewood species that also had less bio-char and less parent material ash (Tables 3 and 4). *E. globulus* was the highest bio-oil yielder (43%, w/w) and locally preferred for firewood (Table 3). This was confirmed by other studies, as the amount of bio-oil yield of pyrolysis of *E. globulus* and pine wood was 75 (%, w/w) at 500 °C, at a rate of 1000 °C s^{-1} and 1 s vapor residence time (Granada et al. 2013). Maximum bio-oil yield of about 70–80% on dry basis of woody biomass pyrolysis was obtained at 480–530 °C from previous other studies. The temperature limit of 600 °C used in the present study was in line with other studies for lignocellulosic biomass (Amutio et al. 2012). The overall yield of bio-oil was similar with other studies of different species; however, different tree species could require different levels of temperature, rate of heat, and vapor residence time (Okoroigwe et al. 2015), which require further study for each species.

Bio-oil and bio-char production of woody biomass depends on the structural components, cellulose, hemicellulose, and lignin. According to Akhtar and Amin (2011), higher content of cellulose and hemicelluloses favors formation of bio-oil; and higher lignin content favors formation of bio-char. *E. globulus* that produced the highest bio-oil (Table 3) contained 50% (w/w) cellulose (Dmitry and Neto 2007), than *G. robusta,* 46.38% (w/w) (Madan and Roymoulik 1990). However, the cellulose content of *C. edulis* in Yishak (2014) was 59% but had low bio-oil amount (25.83%, w/w) in the present study and requires further study. The ash content of bio-oil of the woody biomasses was lower than other studies (Table 4), which could be because of the lower ash content of the parent firewood (Table 4). The highest bio-char obtained from leaf residue of *C. edulis* (Table 3) could be attributed to the highest extractive and lignin content, about 31.5% (Yishak 2014). *G. robusta* was a recently introduced species when compared with a century old *E. globulus* and *E. camaldulensis* in the study area. Most of the fuel characteristics of *G. robusta* bio-oil were comparable to *E. globulus*, and therefore, it can be used as firewood plantation tree species in the area.

The calorific value of the bio-oil in the present study was relatively higher than other studies (Oduor and Githiomi 2013) (Fig. 5), which might be because of centrifuging the bio-oil with distilled water and subsequent dehydration. Better calorific value was obtained from the bio-oil than the parent firewood. Depending on the level of technology, the same wood from a tree species can be used in three different forms of biomass fuels including firewood, bio-oil (bio-char), or charcoal. The processing from firewood to bio-oil diversifies income source. The production processes of bio-char and charcoal have similar pyrolysis technique with similar calorific value, about 28–30 MJ kg^{-1} (Namaalwa et al. 2007). Therefore, in order to reduce the deforestation rate, conversion to bio-oil and bio-char increases the efficiency of biomass use and improves the adaptation to climate change.

In the bio-char, the amount of fixed carbon and volatile matter at moisture content 1.3–7.74 (%, w/w) (Fig. 6) were comparable to other studies. Volatile matter of less than 20% and bio-char fixed carbon content over 50% from all of the 17 species (Fig. 6) obtained in the present study indicated the absence of labile carbon and their potential of long time carbon storage (Awad et al. 2012).

Since GHG emissions from bio-oil and firewood are taken up by the re-growing trees, there is no net emission except end use emission as indoor pollutant smoke. The end use emission as indoor pollution of bio-oil is lower than the firewood combustion because in bio-oil non-condensable gases are removed in manufacturing areas. Therefore, the use of bio-oil is safer than the direct use of firewood because of the reduction of smoke (Demirbas 2004).

The carbon storage in bio-char is a synergy to afforestation and to reduce climate change (Lehmann et al. 2006). The bio-oil and bio-char production technology could also use cheap sources of feedstock like organic wastes to reduce pressure on forests, to dispose waste and retard emission. Moreover, the reduction in forest biomass utilization improves water resources availability, secures clean air to breathe, and generates income by selling the forest products so that climate change adaptation could be improved.

Conclusion and Recommendation

The pyrolysis of woody biomasses of different tree species produced different quantity and quality of bio-oil and bio-char yield. Since the bio-oil and bio-char yield of *A. seyal, D. angustifolia, E. schimperi, E. globulus, C. equisetifolia*, and *G. robusta* was over 62% (w/w) of the parent firewood biomass used in pyrolysis, these can be selected for plantation development and climate change adaptation. Centrifuging pyrolysis oil with distilled water and subsequent dehydration resulted in increased calorific value to 33 MJ kg^{-1} in *E. globulus*. The production and simultaneous use of bio-oil yield (21.1–42.87%, w/w) and bio-char yield (23.233–36.40%, w/w) for household cooking energy and for carbon storage, respectively, instead of firewood reduced end use emission by 71.48–118.06% in each household of lowland to highland PA of the studied area, which could increase adaptation to climate change by reducing the cost of indoor pollution. Therefore, pyrolysis of biomass generally reduces wood wastage, creates jobs, and provides organic carbon as adaptation to climate change. In order to reduce transport cost of low density and high volume biomass, and to reduce the number of transportation vehicles, it is important to establish small-scale biomass pyrolysis firm in electrified parts of Ethiopia that produce bio-oil and bio-char from woody and non-woody organic wastes to diversify income and to increase the adaptive capacity of the rural people to climate change. Moreover, carbon storage potential of bio-char can be used to improve soil fertility in rural areas as a means of climate change adaptation and mitigation. The tree/shrub species used for making charcoal in Ethiopia are mainly slow growing indigenous species existing in natural forest, and there should be a guiding policy to plant tree species like *A. seyal, D. angustifolia, E. schimperi, E. globulus, C. equisetifolia*, and *G. robusta* for charcoal making and to strengthen climate change adaptation.

Acknowledgments The research was financially supported by Hawassa University, Wondo Genet College of Forestry and Natural Resources, Central Ethiopia Environment and Forestry Research Center, and Association of African Universities Small Grant for Post Graduate Theses and Dissertations.

References

Akhtar J, Amin N (2011) A review on process conditions for optimum bio-oil yield in hydrother-mal- liquefaction of biomass. Renew Sust Energ Rev 15:1615–1624
Aliyu B, Agnew B, Douglas S (2010) *Croton megalocarpus* (Musine) seeds as a potential source of Bio-diesel. Biomass Bioenergy 34(10):1495–1499. https://doi.org/10.1016/j.biombioe.2010.04.026

Amutio M, Lopez G, Artetxe M, Elordi G, Olazar M, Bilbao J (2012) Influence of temperature on biomass pyrolysis in a conical spouted bed reactor. Resour Conserv Recycl 59:23–31

Arias BR, Pevida CG, Fermoso JD, Plaza MG, Rubiera FG, Pis-Martinez JJ (2008) Influence of torrefaction on the grindability and reactivity of woody biomass. Fuel Process Technol 89 (2):169–175

ASTM (American Society for Testing and Materials) (1989) Annual book of ASTM standards, petroleum products, lubricants and fossil fuels. Sect 5, v 5 Gaseous fuels, coal and coke. ASTM, West Conshohocken. PCN-01-050589-13

Awad YM, Blagodatskaya E, Ok YS, Kuzyakov YY (2012) Effects of polyacrylamide, biopolymer, and biochar on decomposition of soil organic matter and plant residues as determined by ^{14}C and enzyme activities. Eur J Soil Biol 48:1–10

Bekele TA (2007) In: Tengnäs B, Kelbesa E, Demissew S, Maundu P (eds) Useful trees and shrubs of Ethiopia: identification, propagation and management for 17 agroclimatic zones. RELMA in ICRAF Project World Agroforestry Centre, Nairobi, p 559

Bird N, Cowie A, Cherubini F, Jungmeier G (2011) Using a life cycle assessment approach to estimate the net greenhouse gas emissions of bioenergy. IEA bioenergy P 20, www. ieabioenergy.com. Accessed 17 Oct 2019

Brown R (2009) Bio-char production technology. In: Lehmann J, Joseph S (eds) Bio-char for environmental management: science and technology. Earthscan, London, pp 127–146

Chidumayo EN, Gumbo DJ (2013) The environmental impacts of charcoal production in tropical ecosystems of the world. A synthesis. Energy Sustain Dev 17:86–94

Cordeiro LG (2011) Characterização e viabilidade econômica de bagaço de malte oriundo de cervejarias para fins energéticos João pessoa, 120 f

Demirbas A (2004) The influence of temperature on the yields of compounds existing in bio-oils obtained from biomass samples via pyrolysis. Fuel Process Technol 88:591–597

Dmitry VE, Neto CP (2007) Recent advances in Eucalyptus wood chemistry: structural features through the prism of technological response. https://www.researchgate.net/publication/237632161. Accessed on 16 Nov 2017

Elliot DC (2012) Biomass pyrolysis to liquid fuels in the US 2G 2020, Biofuel Seminar, Helsinki, Finland, August 30, 2012

Eze JN, Ibrahim PA, Tiamiyu SA, Alfa M (2020) Assessment of drought occurrences and its implications on agriculture in Niger State, Nigeria. Discov Agric 6(15):1–10

FAO (Food and Agricultural Organization of the United Nations) (2017) The charcoal transition: greening the charcoal value chain to mitigate climate change and improve local livelihoods, by van Dam Rome J. FAO

Forest Products Laboratory (2004) Fuel value calculator. USDA Forest Service, Forest Products Laboratory, Pellet Fuels Institute, Madison. Available at http://www.fpl.fs.fed.us. Accessed on 17 July 2017

Garcia-Perez M (2008) The formation of polyaromatic and dioxins during pyrolysis: a review of the literature with description of biomass composition, fast pyrolysis technologies and thermochemical reactions. Washington State University. http://www.pacificbiomass.org/documents/theformationofpolyaromatihydrocarbonsanddioxinsduringpyrolysis.pdf. Accessed on 17 July 2017

Granada E, Míguez JL, Febrero L, Collazo J, Eguía P (2013) Development of an experimental technique for oil recovery during biomass pyrolysis. Renew Energy 60:179–184

IPCC (Intergovernmental Panel on Climate Change) (2006) Guidelines for national greenhouse gas inventories, and Emission Factor Data Base (EFDB) version from November 2016, number of emission factors: 16877. Chapter 1, V 2: Energy, Tables 14, 25 & 29

Jindo K, Mizumoto H, Sawada Y, Sanchez-Monedero MA, Sonoki T (2014) Physical and chemical characterization of biochars derived from different agricultural residues. Biogeosciences 11:6613–6621

Kannan P, Arunachalam P, Prabukumar G, Govindaraj M (2013) Bio-char an alternate option for crop residues and solid waste disposal and climate change mitigation. Afr J Agric Res 8 (21):2403–2412. https://doi.org/10.5897/AJAR12.2083

Krajnc N, Prislan P, Jemec T, Triplat M (2014) Development of biomass trade and logistics centres for sustainable mobilisation of local wood biomass resources – Biomass Trade Centre II:

publishable report Ljubljana: Gozdarski inštitut Slovenije, 2014 ilustr http://proforbiomed.eu/sites/default/files/14%20-%20Environmental%20impact.pdf. Accessed on 21 Dec 2016

Lehmann J, Gaunt J, Rondon M (2006) Bio-char sequestration in terrestrial ecosystems: a review. Mitig Adapt Strateg Glob Chang 11:403–427

Madan RN, Roymoulik SK (1990) Evaluation of Grevillea robusta as a raw material for the production of dissolving pulp and viscose rayon. Indian J Fibre Text Res 15:180–184

Miftah F, Tsegaye B, Sisay F (2017) Impact of energy consumption on indoor pollution in rural wooden houses of Southern Ethiopia in times of climate change. Glob J Curr Res 4(3):14–26. ISSN: 2320-2920

Namaalwa J, Sankhayan PL, Hofstad O (2007) A dynamic bio-economic model for analyzing deforestation and degradation: an application to woodlands in Uganda. Forest Policy Econ 9:479–495

Oduor NM, Githiomi JK (2013) Fuel-wood energy properties of *Prosopis juliflora* and *Prosopis pallida* grown in Baringo District, Kenya. Afr J Agric Res 8(21):2476–2481. https://doi.org/10.5897/AJAR08.221. http://www.academicjournals.org/AJAR

Okoroigwe EC, Zhenglong LI, Shantanu K et al (2015) Bio-oil yield potential of some tropical wody biomass. J Energy South Afr 26(2):33–41

Roberts KG, Glory BA, Joseph S et al (2010) Life Cycle Assessment of bio-char systems: Estimating the energetic, economic and climate change potential. Environ Sci Technol 44:827–833

Shackley S et al (2012) Biochar, tool for climate change mitigation and soil management. In: Meyers RA (ed) Encyclopedia of sustainability science and technology. Springer, New York

Sheng C, Azevedo JLT (2005) Estimating the higher heating value of biomass fuels from basic analysis data. Biomass Bioenergy 28(5):499–507. https://doi.org/10.1016/j.biombioe.2004.11.008

Shimelis B (2011) Generation, composition and characteristics of urban solid waste in a major Khat producing and marketing area in Eastern Ethiopia. Int J Environ Prot I(5):9–16. Retrieved from http://www.ij-ep.org

Xiu S, Shahbazi A (2012) Bio-oil production and upgrading research: a review. Renew Sust Energ Rev 16:4406–4414. https://doi.org/10.1016/j.rser.2012.04.028

Yishak AY (2014) Liquid fuel production through pyrolysis of Khat and plastic waste mixture. A Thesis Submitted in Partial Fulfillment of the Requirements for Master Degree in Chemical Engineering under Env Engi Addis Ababa Institute of Technology (AAiT), School of Chemical and Bio-Engineering

10

Constraints to Farmers' Choice of Climate Change Adaptation Strategies in Ondo State of Nigeria

George Olanrewaju Ige, Oluwole Matthew Akinnagbe, Olalekan Olamigoke Odefadehan and Opeyemi Peter Ogunbusuyi

Contents

Introduction .
   Background
Description of the Study Area and Methodology

   Limitations of the Study
Results and Discussion
   Socio-Economic Characteristics of the Respondent
   Constraints to Choice of Adaptation Strategies Used by the Respondents
   Relationship Between Some Socioeconomic Characteristic of the Respondents
   and Constraints to Choice of Adaptation Strategies
Conclusions ....
References

Abstract

Nigeria being dependent on rain-fed agriculture and with low level of socio-economic development is highly affected and vulnerable to climate change. It

G. O. Ige (✉) · O. M. Akinnagbe · O. O. Odefadehan
Department of Agricultural Extension and Communication Technology, School of Agriculture and Agricultural Technology, Federal University of Technology, Akure, Nigeria
e-mail: igego@futa.edu.ng; omakinnagbe@futa.edu.ng; olalekanodefadehan@gmail.com

O. P. Ogunbusuyi
Department of Agricultural and Resource Economics, School of Agriculture and Agricultural Technology, Federal University of Technology, Akure, Nigeria
e-mail: ogunbusuyiopeyemi@gmail.com

is crucial for farmers to adapt to the never ending climate change. However, there are constraints to adaptation strategies used by the farmers. This study therefore identified some of the constraints to the farmers' choice of climate change adaptation strategies in Ondo State, Nigeria. A multistage sampling procedure was used in selecting one hundred and sixty respondents for the study. Data collected with a well-structured interview schedule were analyzed using frequency, percentage, and mean statistic, while Pearson Product Moment Correlation was used to test hypothesis. Crops competing for nutrient, inadequate access to climate information, inadequate finance, scarcity of labor, and inadequate farm input supplies were among the major constraints to choice of climate change adaptation strategies used by the respondents. The study recommended that weather forecast information should be published and made available to the farmers through agricultural extension agents. Training on how to improve mixed cropping technique and avoid vulnerability should be pursued.

Keywords

Constraints · Climate change · Adaptation strategies · Farmers · Farm-inputs

Introduction

Background

Climate change is one of the major sources of risk in agriculture (De and Badosa 2015). In sub-Saharan Africa, it is set to hit the sector severely and cause suffering, particularly for smallholder farmers. The potential of agriculture to generate a more pro-poor growth process depends on the creation of new market opportunities that most benefit the rural poor (Culas and Hanjra 2011). This could be attributed to the fact that climate change affects the two most important direct agricultural production inputs: precipitation and rainfall (Deschenes and Greenstone 2006). Climate change also indirectly affects agriculture by influencing emergence and distribution of crop pests and livestock diseases, exacerbating the frequency and distribution of adverse weather conditions, reducing water supplies for irrigation, and enhancing severity of soil erosion (IPCC 1998).

Nigeria, which is dependent on rain-fed agriculture together with low level of socio economic development, is highly affected and vulnerable to climate change. Thus, a better understanding of farmers' concerns and their perception of climate change is crucial to design effective policies for supporting successful adaptation of the agricultural sector (Kumar and Sidana 2018). Adaptation helps farmers achieve their food, income, and livelihood security objectives in the face of changing climatic and socioeconomic conditions, including climate variability, extreme weather conditions such as droughts and floods, and volatile short-term changes in local and large-scale markets (Kandlinkar and Risbey 2000). Farmers can reduce the potential damage by making tactical responses to these changes.

Adaptation strategies which are applicable to climate change are not exhaustive. These strategies vary with scope, approach, purpose, use, application, and timing. The choice of adaptation methods by farmers depends on various social, economic, and environmental factors (Bryan et al. 2013). Adaptation is therefore important for finding ways to help farmers adapt in the rural economies of Africa by providing effective adaptation strategies such as the use of irrigation facilities, improved and resistant crop varieties, modern farm mechanization among others in combating the adverse effect of climate change.

However, there are constraints to adaptation strategies of the farmers, which vary across countries. Satishkumar et al. (2013) categorized constraints faced by farmers in India into personal, institutional, and technical constraints. Small scale fragmented land holdings, low literacy level, inadequate knowledge of how to cope or build resilience, and traditional beliefs were categorized as personal constraints; poor access to extension services, poor access to information sources, and nonavailability of institutional credit were categorized as institutional constraints, while non-availability of drought tolerant varieties, lack of access to weather forecasting information, dependent on monsoon irrigation, high cost of irrigation facilities, shifting of cropping pattern, and lack of technical know-how were categorized as technical constraints.

Fagariba et al. (2018) ranked the constraints affecting farmers' efforts to curb the impact of climate change in Northern Ghana. The study reported unpredictable weather conditions to be the most serious constraint to climate change adaptation strategies. Inadequate government support, lack of access to weather information, land tenure issues, high cost of input, inadequate extension officers, lack of formal education, and poor soil fertility were the other constraints. Otitoju and Enete (2016) grouped the constraints to the use of adaptation strategies among food crop farmers in South-western Nigeria into four. The groups include public, institutional, and labor constraints; land, neighborhood norms, and religious beliefs constraints; high cost of inputs, technological and poor information on early warning systems constraints; and far farm distance, poor access to climate change adaptation information, off-farm job, and credit constraints. Evers and Pathirana (2018) submit that greater communication, cooperation, and coordination of both policy and scientific work that cut across both disciplinary and geographical boundaries are needed to help reduce uncertainties in future climate projections and inform adaptation decisions.

As site-specific issues require site-specific knowledge, it is very important, therefore, to clearly understand what is happening at community level, because farmers are the most climate vulnerable group. In the absence of such location specific studies, it is difficult to fine tune interventions geared towards alleviating the constraints faced by the local farmers. It is on this note that this study seeks to know the socio economic characteristics of the farmers' and also identify the constraints to farmers' choices of climate change adaptation strategies in the study area.

Description of the Study Area and Methodology

Study Area

The study was carried out in Ondo State, one of the 36 states in Nigeria, and a member of the Niger Delta commission (NDDC) in Nigeria. Ondo State was created on third of February 1976 from the former western state. It originally included what is now Ekiti state, before the separation in 1996 (ODGS 2016). It is bounded in the North West by Ekiti State, west central by Osun State, South East by Ogun State, South East by Edo and Delta states, and the South by the Atlantic Ocean. The state lies between latitude 5° 45 and 8° 15^1 north and longitude 4° 45^1 and 6° 05 ^{1} east. Its land area is about 15,500 square kilometers (ODGS 2016).

Ondo State's climate is of two distinct seasons (wet and dry season). In the south, the mean monthly temperature is 27 °C with a mean monthly range of 2 °C, while mean relative humidity is over 75%. However, in the northern part of the state, the mean monthly temperature and its range are about 30 °C and 6 °C, respectively. The mean monthly relative humidity is less than 70%. Also, the mean annual total rain fall exceed 2000 mm. However, in the northern part, there is marked dry season from November to march when little or no rain fall, hence, the total annual rain fall in the North drops considerably to about 1800 mm (Fasina et al. 2016; ODSG 2016).

The natural vegetation is the high forest, composed of many varieties of hardwood timber such as *Melicia excelsa, Antaris toxicaria, Lophira alata, Terminalia superba, and Symphonia globulinfera.* An important aspect of vegetation of the state is the prevalence of tree crops. The major tree crops include cocoa, kola, rubber, coffee, oil palms, fruits, and citrus (Institute of Food Security, Environmental Resource and Agricultural Research (IFSERAR) 2010). Ondo state is predominantly an agricultural state with over 60% of its labor force deriving their income from farming. It is the leading Cocoa producing state in Nigeria. Other agricultural products include yam, cassava, cocoyam, plantain; thus, the state can support the cultivation of a large variety of crops. It also has the longest coastline in the Country which favor fishing activities in the riverine areas (IFSERAR 2010). The population of the study includes all arable crop farmers in Ondo State, Nigeria.

Research Methods

Study Site Selection and Sampling Methods
A multistage sampling procedure was used in selecting the respondents for this study. In the first stage, two out of the three senatorial districts in the state were randomly selected. The selected senatorial districts are Ondo Central and Ondo North. The second stage involved a random selection of two (2) Local Government Areas (LGAs) from each of the two selected senatorial districts. The selected LGAs

are: Akure South, Ifedore, Ose, and Owo LGAs. The third stage involved purposive selection of four (4) villages, based on the predominance of arable crop farming in the LGAs, making a total of sixteen (16) villages. In the final stage, ten arable crop farmers were randomly selected. Thus, a total of one hundred and sixty (160) respondents constituted the sample size for the study. Data were collected with the aid of a well-structured interview schedule.

Data Source and Collection Method

Data Analysis
Data analytical techniques that were used in achieving the objectives for this study are descriptive statistics. They include frequency counts, percentage and mean.

The socio-economic characteristics of the farmers were analyzed using frequency, percentage, and mean statistic. To identify the constraints to choice of adaptation strategies used by the respondents, a list of possible constraints were made available for the farmers to tick on a three-point Likert- type scale of major constraint, minor constraint, and not a constraint, with the value of 2, 1, and 0, respectively. These values were added up to 3 and divided by 3 to get a mean value of 1. Any variables with mean value greater than or equals to 1 imply that they are constraints to climate change adaptation, while mean value of less than 1 implies that they are not a constraint. Any variable with a mean value of greater than or equals to 1.5 are major constraints. The responses were analyzed using mean statistic.

Hypothesis H_0: There is no significant relationship between some socio economic characteristics of the respondents and constraints to choice of adaptation strategies used by the respondents. This was tested using the Pearson Product Moment Correlation (PPMC).

Limitations of the Study

The study covered only four (4) LGAs out of the eighteen (18) LGAs in the state. Only one hundred and sixty respondents out of several hundred thousands of arable crop farmers in the state were included in the study. This was due to the paucity of funds to carry out the study. A funded research can cover a larger sample for a similar study in the future. Furthermore, access to some farmers was very difficult due to bad access roads, which is a feature of many rural communities in Nigeria. Responses gotten from the respondents were based on memory recall which may not be perfectly and correctly stated. The effects of this limitation were minimized by repeating some questions but reframed, so as to confirm the response to the previous question. Uncooperative attitude of some farmers in terms of giving accurate information about themselves was another limitation. These limitations propelled the researchers to put extra effort to collect reliable data, which will be affected by the identified limiting factors.

Results and Discussion

Socio-Economic Characteristics of the Respondent

The distribution of respondents by their socio-economic characteristics is presented in Table 1. It was revealed that 38.8% were between 50 and 59 years of age, 24.5% were between 40 and 49 years, 20.7% were between 60 and 69 years, while 10.5% of the respondents were above 70 years of age. Also, 4.9% of the respondents were between 30 and 39 years, while only 0.6% of the respondents were less than 30 years of age. It was further revealed that the average age of the respondents was 54.6 years. This is an indication that arable crop farmers in the study area are fairly old, although they are still within the active age with the strength and vigor to carry out all the laborious activities involved in agricultural production. This agreed with the findings of Oluwasusi and Tijani (2013), who reported an average age of 53.9 years in a study conducted in Ekiti State, Nigeria.

Farming in the study area was dominated by the males (62.5%), while the female respondents accounted for 37.5% of the respondents. This might be attributed to the laborious nature of all the activities involved in arable crop production/farming in the study area. However, the females were believed to be more involved in the processing and marketing of farm by-products. This agreed with the findings of Ibitoye et al. (2014) who reported that the males dominated agricultural production in a study on the constraints to climate variability adaption among arable crop farmers in Ekiti State, Nigeria.

It was further revealed that majority (87.5%) of the respondents were married, while only 1.9% of them were still single, 6.3% of the respondents were divorced, while 4.4% of them were separated. The fact that majority of the farmers in the study area were married might be a reflection of the strong moral values attached to marriage institution in the study area. This also agreed with the findings of Ibitoye (2012) and Oguntade et al. (2014) who reported in their separate studies that arable crop farmers that are married dominated arable crop farming activities in Nigeria. Majority (62.5%) of the arable crop farmers in the study area were Christians, 32.5% of the respondents were Muslims, while 5.0% of the respondents practiced traditional religion.

Average household size of the respondents was 7 people. It was revealed that majority (71.2%) of the farming households had between 5 and 9 people, while 16.3% had less than 5 people in their household. However, 12.5% of the respondents had 10 people and above in their respective households. The average household size of about 7 people implied that the household size among farming households in the study area is fairly large. However, this is expected to be a very good source of labor (family labor) for their farming activities and it is expected to enhance their level of productivity.

It was observed that 28.1% of the respondents had no formal education, 21.9% completed primary school education, while 15.6% of them completed secondary school education. It was also revealed that 11.3% attempted primary school education, 8.1% of the respondents acquired adult education, 8.1% of the respondents

Table 1 Socio-economic characteristics of the respondents

Socioeconomic characteristics	Frequency	Percentage (%)	Mean ($\overline{X}$)
Age (years)			54.6
<30 years	1	0.6	
30–39 years	8	4.9	
40–49 years	39	24.5	
50–59 years	62	38.8	
60–69 years	33	20.7	
70 years and above	17	10.5	
Sex			
Female	60	37.5	
Male	100	62.5	
Marital status			
Single	3	1.9	
Married	140	87.5	
Widowed	10	6.3	
Divorced	7	4.4	
Religion			
Christian	100	62.5	
Islam	52	32.5	
Traditional	8	5.0	
Household size (number)			7
<5 persons	26	16.3	
5–9 persons	114	71.2	
10 persons and above	20	12.5	
Level of education			
No formal education	45	28.1	
Adult education	13	8.1	
Attempted primary school	18	11.3	
Completed primary school	35	21.9	
Attempted secondary school	13	8.1	
Completed secondary school	25	15.6	
Attempted tertiary education	1	0.6	
Completed tertiary education	10	6.3	
Farming experience (years)			27.5
<10 years	16	10.1	
10–19 years	29	18.1	
20–29 years	41	25.5	
30 years and above	74	46.3	
Farm size (hectares)			0.91
<1	114	71.4	
1–1.99	28	17.5	
2–4.99	16	10	
≥5	2	1.1	

(continued)

Table 1 (continued)

Socioeconomic characteristics	Frequency	Percentage (%)	Mean $(\overline{X})$
Annual income (₦)			318,595.91
<100,000	14	8.7	
100,000–499,999	113	70.6	
500,000–999,999	28	17.5	
1,000,000 and above	5	3.2	

Source: Field Survey, 2017

attempted secondary school education, while 6.3% of them completed tertiary education. However, 0.6% of the respondents attempted education at tertiary level, without completing it. The fact that only 15.6% of the total respondents completed secondary school education and that only 6.3% of them completed tertiary education is an indication that the level of education among the farmers in the study area is low, in agreement with the findings of Oluwasusi and Tijani (2013). This is believed to have an effect on their level of awareness about climate change effects and also on their level of adoption on strategies in mitigating its effects.

It was further revealed that respondents that had farming experience of 30 years and above constituted 46.3% of the total respondents sampled for this study. However, 25.5% of the respondents were reported to have between 20 and 29 years of farming experience, 18.1% had between 10 and 19 years of farming experience, while only 10.1% of the total respondents had less than 10 years of farming experience. It was also revealed that the average years of farming experience among the arable crop farmers in the study area was 27.5 years. This is an indication that respondents are well experienced in farming activities in the study area. This is expected to enhance the knowledge of the farmers in how to select the appropriate measures to mitigate and cope with the effect of climate change in the study area, in agreement with the findings of Oguntade et al. (2014).

Majority (71.4%) of the respondents had less than one hectare of farm land, 17.5% have between 1–1.99 ha, while 10.0% of the respondents had a farm size of between 2 and 4.99 ha. However, 1.1% of the respondents had a farm size of five hectares and above. The average farm size was approximately one hectare. This suggests that respondents in the study area are small holder arable crop farmers, who practice farming on subsistence basis. This finding is in agreement with the findings of Babatunde (2008), who reported that majority of arable crop farmers in Nigeria are small holder farmers who practice subsistence farming on a small piece of land.

Majority of the respondents (70.6%) earned between ₦100,000 and ₦499,999 per annum, 17.5% of them earned between ₦500,000 and ₦999,999 per annum, while 8.7% of the respondents earned less than ₦100,000 per annum. However, 3.2% of the respondents earned above ₦1,000,000 per annum. The average annual income of the respondents was also revealed to be ₦318,595.91. This suggests that the average annual income of the respondents was fairly low. This could, however, be attributed to the fact that the arable crop farmers in the study are operating on a subsistence scale, with majority of them farming on less than one hectare of land as reported earlier. The low income earned by the arable crop farmers might equally be due to low output that usually occurs as a result of the adverse effects of climate change. This is in agreement with the findings of Falola et al. (2012); also, low income earned by the arable crop farmers may also be as a result of extra costs incurred by the arable crop farmers in an attempt to mitigate the adverse effects of climate change.

Constraints to Choice of Adaptation Strategies Used by the Respondents

Table 2 presents the constraints to choice of adaptation strategies used by the respondents in the study area. It was revealed that inadequate finance ($\overline{X} = 1.8$), scarcity of labor ($\overline{X} = 1.8$), and inadequate technical know-how knowledge ($\overline{X} = 1.7$) were the major constraints militating against the choice of irrigation as a coping strategy to the effect of climate change in the study area. The problems of inadequate finance would not enable the arable crop farmers to acquire the necessary irrigation equipment and facilities which might be too expensive for them to acquire from the proceeds from their farming activities. Also, inadequate knowledge and technical know-how in operating the modern irrigation equipment will to a greater extent limit the farmers use of irrigation as an adaptation strategy to the effect of climate change in the study area.

It was further revealed that poor agricultural extension services ($\overline{X} = 1.6$), high cost of fertilizer and other inputs ($\overline{X} = 1.5$), as well as pest and diseases ($\overline{X} = 1.6$) were the major constraints militating against the choice of planting of cover crops as a coping strategy to combat the effect of climate change in the study area. This implied that inadequate dissemination of information as a result of poor extension services limits the level of awareness of the farmers on the potentials the planting of cover crops has in reducing the effect of adverse climatic condition (e.g., drought) on arable crops in the study area. This is in conformity with the findings of Satishkumar et al. (2013) Also, high cost of fertilizer and other inputs required in planting and maintaining cover crops constrained some of the arable crop farmers to use the planting of cover crops to combat the adverse effect of climate change in the study area. This is in line with the findings of Otitoju and Enete (2016). High possible incidence of pest and disease infestation associated with the use of cover crops can also be a constraint to the use of cover crop in mitigating the effects of climate change in the study area.

It was further revealed that the major constraints militating against the use of weather forecasting as an adaptation strategy against the effects of climate change included inadequate access to climate information ($\overline{X} = 1.8$). This corroborates the findings of Fagariba et al. (2018). This suggests that inadequacy of climatic information available to the arable crop farmers in terms of accuracy and consistency of metrological information greatly constrained them from choosing weather forecasting as an adaptation strategy against the effects of climate change. Also, the high cost of accessing the weather forecast information as well as the incompatibility of such information constrained the farmers in adopting weather forecasting as a reliable means of coping with the effect of climate change in the study area.

Table 2 Constraints to choice of adaptation strategies used by the respondents

Constraints to adaptation strategies	Major constraints (2)	Minor constraints (1)	Not a constraint (0)	Mean $(\bar{X})$	Std. dev.
Irrigation					
Inadequate finance	38 (86.4)	4 (9.1)	2 (4.5)	1.8*	0.5
Land related issue	35 (79.5)	1 (2.3)	8 (18.2)	1.6*	0.8
Scarcity of labor	37 (84.1)	6 (13.6)	1 (2.3)	1.8*	0.4
Inadequate technical know-how knowledge	36 (81.8)	2 (4.5)	6 (13.6)	1.7*	0.7
Planting of cover crops					
Poor agricultural extension services	38 (77.6)	3 (6.1)	8 (16.3)	1.6*	0.8
Farm distance	30 (61.2)	4 (8.1)	15 (30.7)	1.3	0.4
High cost of fertilizer and other inputs	34 (69.4)	6 (12.2)	9 (18.4)	1.5*	0.8
Pest and diseases	38 (76.0)	4 (8.0)	8 (16.0)	1.6*	0.8
Use of weather forecasting					
Inadequate access to climate information	30 (83.3)	3 (8.3)	3 (8.3)	1.8*	0.6
Cultural incompatibility	26 (72.2)	6 (16.7)	4 (11.1)	1.6*	0.7
It is expensive and depends on technologies	28 (77.8)	2 (5.6)	6 (16.7)	1.6*	0.8
Inconsistent government polices	22 (62.9)	5 (14.3)	8 (22.9)	1.4	0.6
Use of resistant crop varieties					
Availability of the varieties	44 (88.0)	1 (2.0)	5 (10.0)	1.8*	0.6
Cost of the varieties	41 (83.7)	–	8 (16.3)	1.7*	4.1
Farm distance	32 (65.3)	6 (12.2)	11 (22.4)	1.4	0.8
Off farm job and credits	34 (70.8)	5 (10.4)	9 (18.8)	1.5*	0.8
Change in planting date					
Poor agricultural program and service delivery	50 (82.0)	4 (6.6)	7 (11.5)	1.7*	0.7
Neighborhood norms and religious believes	29 (47.5)	11 (18.5)	21 (34.4)	1.1	0.9
Reduction in output	41 (67.2)	9 (14.8)	11 (18.0)	1.5*	0.8
Inadequate access to hybrid seeds	50 (82.0)	6 (9.8)	5 (8.2)	1.7*	0.6
Change in harvest date					
Inaccurate agro-metrological information	19 (79.2)	1 (4.2)	4 (16.7)	1.6*	0.8
Time of dissemination of agro-meteorological information	17 (70.8)	3 (12.5)	4 (16.7)	1.5*	0.8
Poor storage and processing facilities	16 (66.7)	2 (8.3)	6 (25.0)	1.4	0.9
High cost of farm operations	14 (58.3)	2 (8.3)	8 (33.3)	1.3	0.9

(continued)

Table 2 (continued)

Constraints to adaptation strategies	Major constraints (2)	Minor constraints (1)	Not a constraint (0)	Mean ($\overline{X}$)	Std. dev.
Use of sand bags by river banks					
High cost of labor	9 (100.0)	–	–	2.0*	0.0
Inadequate finance and credit facilities	9 (100.0)	–	–	2.0*	0.0
Drudgery of making the sand bags	8 (88.9)	1 (11.1)	–	1.9*	0.3
Longer time in making the sand bags	8 (88.9)	–	1 (11.1)	1.8*	0.7
Use of mulching materials					
Over grazing of the land	25 (39.1)	6 (9.4)	33 (51.6)	0.9	0.5
Urbanization	21 (31.8)	9 (13.6)	36 (54.5)	0.8	0.9
High cost of labor in applying the mulching material	51 (73.3)	11 (16.7)	4 (6.1)	1.7*	0.6
Durability of the mulching materials	29 (44.6)	21 (32.3)	15 (23.1)	1.2	0.8
Mixed cropping					
Vulnerability to pest and diseases	42 (52.5)	11 (13.8)	27 (33.8)	1.2	0.9
Crops competing for nutrient	48 (60.8)	17 (21.5)	14 (17.7)	1.4	0.8
Depletion of the soil nutrient	38 (48.1)	17 (21.5)	24 (30.4)	1.2	0.9
Poor extension delivery services	35 (44.9)	19 (24.4)	24 (30.8)	1.1	0.9
Crop rotation					
Life span of the crop grown	11 (47.8)	6 (26.1)	6 (26.1)	1.2	0.9
Inadequate farm input supplies	15 (68.2)	6 (27.3)	1 (4.5)	1.6*	0.6
Depletion of the soil nutrient	8 (36.4)	3 (13.6)	11 (50.0)	0.9	0.9
Nonavailability of labor	6 (27.3)	8 (36.4)	8 (36.4)	0.9	0.8
Use of intercropping					
Complication of crop management and harvesting	7 (43.8)	3 (23.1)	3 (23.1)	1.3	0.9
Problem of separating seed crop yield	4 (30.8)	5 (38.5)	4 (30.8)	1.0	0.8
Community customs and laws	3 (25.0)	2 (16.7)	7 (58.3)	0.7	0.9
Nonfarm income diversification					
Government policies due to taxes	1 (8.3)	3 (25.0)	8 (66.7)	0.4	0.9
Inadequate finance and credit	23 (71.9)	1 (3.1)	8 (25.0)	1.5*	0.9
Poor skills	22 (66.7)	4 (3.0)	7 (24.2)	1.5*	0.8
Low wages and poor condition of work	24 (72.7)	1 (3.0)	8 (24.2)	1.5*	0.9

Key: 0–0.9 = Not a Constraint; 1.0–1.4 = Minor Constraint; 1.5 and above = Major Constraint
*denotes major constraints
Note: Values in parenthesis are Percentages

It was further revealed in this study that the major factors militating against the use of resistant crop varieties as an adaptation strategy against the effects of climate change in the study area included nonavailability of the disease resistant varieties ($\overline{X} = 1.8$), high cost of the varieties ($\overline{X} = 1.7$), as well as off farm job and credits of the farmer ($\overline{X} = 1.5$). The high cost, coupled with the unavailability of improved varieties of arable crops that are resistant to pest and diseases, is a major constraint preventing the arable crop farmers from adopting the use of resistant crop varieties to cope with the effect of climate change in the study area. Otitoju and Enete (2016) reported a similar finding.

Poor agricultural program and service delivery ($\overline{X} = 1.7$), inadequate access to hybrid seeds ($\overline{X} = 1.7$), and reduction in output ($\overline{X} = 1.5$) were the major constraints militating against the use of change in planting date as an adaptation strategy against the effects of climate change in the study area. Also, inaccurate agro-metrological infor-

mation ($\overline{X} = 1.6$), time of dissemination of agro-metrological information ($X = 1.5$), and poor storage and processing facilities ($\overline{X} = 1.4$) were the major constraints militating against the use of change in harvest date as an adaptation strategy to the effects of climate change in the study area. This implied that inaccurate and unseemliness of agro-metrological information greatly constrains the arable crop farmers in the adoption of both change in planting and harvest dates as a coping strategy in combating the effects of climate change in the study area.

It was also observed that major constraints that militated against the use of sand bags by river banks as an adaptation strategy to the effects of climate change in the study area included high cost of labor ($\overline{X} = 2.0$), inadequate finance and credit facilities ($\overline{X} = 2.0$), and drudgery of making the sand bags ($\overline{X} = 1.9$). This implied that the use of sand bags as an adaptation strategy might be very expensive for the arable crop farmers to adopt. Therefore, with the limited finance available to the arable crop farmers, it might be difficult to adopt the use of sand bags as an adaptation strategy to combat the effect of climate change in the study area. Also, drudgery of making the sand bags might be a discouraging factor preventing the farmers from adopting the use of sand bags as an adaptation strategy to combat the effect of climate change in the study area.

Furthermore, high cost of labor in applying the mulching material ($\overline{X} = 1.7$) and the durability of the mulching materials ($\overline{X} = 1.2$) were the major constraining factors militating against the use of mulching materials as an adaptation strategy against the effects of climate change in the study area. This suggests that the higher the cost of labor in applying the mulching materials, the lesser the adoption of mulching materials as an adaptation strategy in combating the effect of climate change in the study area. It was further observed that the constraints militating against the use of mixed cropping system as an adaptation strategy in the study area were minor ones. They included crops competing for nutrient ($\overline{X} = 1.4$), vulnerability to pest and diseases ($\overline{X} = 1.2$), and depletion of the soil nutrient ($\overline{X} = 1.2$). This might be attributed to the fact that the practice of mixed cropping increase the vulnerability of the crops planted to different pests and diseases. Also, practicing mixed cropping will make the different crops compete for the limited available nutrients, and ultimately this will lead to quick depletion of soil nutrients.

Table 3 PPMC Correlation between some socioeconomic characteristic of the respondent and constraints to choice of adaptation strategies

Variable	r-value	p-value	Decision
Age	−0.028	0.729	Not significant
Household size	0.122	0.128	Not significant
Level of education	−0.074	0.354	Not significant
Farming experience (years)	0.126	0.116	Not significant
Farm size (plots)	0.170[a]	0.034	Significant

[a]Correlation is significant at the 0.05 level (2-tailed)

Furthermore, the major constraining factor against the use of crop rotation as an adaptation strategy to the effects of climate change was observed to be inadequate farm input supplies ($\overline{X} = 1.6$), while the life span of the crop grown ($\overline{X} = 1.2$) was a minor constraint. Moreover, there were only minor constraining factors against the use of intercropping as an adaptation strategy to the effects of climate change in the study area. They include complication of crop management and harvesting ($\overline{X} = 1.3$) and problem of separating seed crop yield ($\overline{X} = 1.0$). However, inadequate finance and credit ($\overline{X} = 1.5$), poor skills ($\overline{X} = 1.5$), and low wages and poor condition of work ($\overline{X} = 1.5$) were the major constraints preventing the arable crop farmers from choosing nonfarm income diversification as an adaptation strategy against the effects of climate change in the study area.

Relationship Between Some Socioeconomic Characteristic of the Respondents and Constraints to Choice of Adaptation Strategies

Table 3 presents the result of the null hypothesis which states that "There is no significant relationship between some socioeconomic characteristic of the respondent and the constraints to choice of adaptation strategies used by the respondent." It was revealed that the farm size of the respondent was the only socioeconomic characteristic that had a strong positive correlation (r = 0.170; p = 0.034) with the constraints to choice of climate change adaptation strategies used by the respondent in the study area. This suggests that constraints to choice of climate change adaptation strategies increases with increase in the farm size, implying that respondents with larger arable crop farms were faced with more constraints to choice of climate change adaptation strategies in the study area compared with those with smaller farm size.

Conclusions

From the findings, it can be concluded that many of the farmers, who happened to be males, fairly old, and are arable crop farmers. The major constraints to the various adaptation strategies were inadequate finance, scarcity of labor, poor agricultural extension services, inadequate access to climate information, nonavailability of resistant varieties, poor agricultural program, inaccurate agro-meteorological information, high cost of labor, low wages and poor condition of work, complication of crop management and harvesting, inadequate farm input supplies, and crops competing for nutrient. These constraints have hindered arable crop farmers in their choice of the identified adaptation strategies in coping with the adverse effect of climate change; it is also reduced farmers level of food production in a bid to ensure food security in the society and also elevate their own economic standard. Hence, respondents with larger arable crop farms were faced with more constraints to choice of climate change adaptation strategies in the study area compared with those with smaller farm size.

It is recommended that weather forecast information should be published and be made available to the farmers for them to be aware of situations to be prepared for at the beginning of each production season through agricultural extension agents. Also, training on how to improve the mixed cropping technique and avoid vulnerability should be pursued. This could also avert crop failure.

References

Babatunde RO (2008) Income inequality in rural Nigeria; evidence from farming household data. Aust J Basic Appl Sci 2(1):134–135

Bryan E, Ringler C, Okoba B, Roncoli C, Silvestri S, Herrero M (2013) Adapting agriculture to climate change in Kenya: household strategies and determinants. J Environ Manage 114:26–35

Culas RJ, Hanjra MA (2011) Some explorations into Zambia's post-independent policies for food security and poverty reduction. In: Contreras LM (ed) Agricultural policies: new developments. Nova Science Publishers, Hauppauge

De UK, Badosa K (2015) Crop diversification in Assam and use of modern inputs under changing climate conditions. Journal of Climatology and Weather Forecasting 2(2):1–14

Deschenes O, Greenstone M (2006) The economic impacts of climate change: evidence from agricultural output and random fluctuations in weather. Am Econ Rev 97(1):354–385

Evers J, Pathirana A (2018) Adaptation to climate change in the Mekong River basin: introduction to the special issue. Clim Change 149:1–11

Fagariba CJ, Song S, Baoro SKGS (2018) Climate change adaptation strategies and constraints in northern Ghana: evidence of farmers in Sissala West District. J Sustain 10(5):1–18

Falola A, Fakayode SB, Akangbe JA, Ibrahim HK (2012) Climate change mitigation activities and determinants in the Rural Guinea Savannah of Nigeria. Sustain Agric Res 1(2). pp16

Fasina OO, Ibitoye AL, Ibitoye O (2016) Determinants of change in work values in rural Nigeria: evidence from Ondo state. J Cult Soc Dev 24:20–27

Ibitoye O (2012) Income inequality among arable crop farming households in rural and urban areas of Ekiti state, Nigeria. Unpublished M. Tech. Thesis in the Department of Agricultural and Resource Economics, Federal University of Technology, Akure

Ibitoye O, Ogunyemi AI, Ajayi JO (2014) Constraints to climate variability adaption among arable crop farmers in Ekiti state, Nigeria. Paper presented at the annual national conference of Nigerian Association of Agricultural Economists (NAAE). Theme: climate change, agriculture and food security in Nigeria. The Federal University of Technology, Akure. 24th – 27th, February, 2014

Institute for Food Security and Environmental Research (IFSERAR) (2010) Impact of climate change on peoples livelihood, wildlife resources and food security in Southwest Nigeria. Annual research report submitted to Institute for Food Security and Environmental Research (IFSERAR), University of Agriculture, Abeokuta

Intergovernmental Panel on Climate Change (IPCC) (1998) The IPCC second assessment: climate change 1998. Cambridge University Press, Cambridge

Kandlinkar M, Risbey J (2000) Agricultural impacts of climate change: if adaptation is the answer, what is the question? Clim Change 45:529–539

Kumar S, Sidana BK (2018) Farmers' perceptions and adaptation strategies to climate change in Punjab agriculture. Indian J Agric Sci 88(10):1573–1581

Oguntade AE, Ibitoye O, Ogunyemi AI (2014) Farmers' perception of climate variability and adaptation strategies for sustainable rice production in Ekiti state, Nigeria. Paper presented at the annual national conference of Nigerian Association of Agricultural Economists (NAAE). Theme: climate change, agriculture and food security in Nigeria. The Federal University of Technology, Akure. 24th – 27th, February, 2014

Oluwasusi JO, Tijani SA (2013) Farmers' adaptation strategies to the effect of climate variation on yam production: a case study in Ekiti state, Nigeria. Agrosearch 13(2):20–31

Ondo State Government (ODSG) (2016). www.ondostate.gov.ng/new/people.php?festId=1. Date accessed: December, 2017

Otitoju MA, Enete AA (2016) Climate change adaptation: uncovering constraints to the use of adaptation strategies among food crop farmers in south-west, Nigeria using principal component analysis (PCA). Cogent Food Agric 2:1. https://doi.org/10.1080/23311932.2016.1178692

Satishkumar N, Tevari P, Singh A (2013) A study on constraints faced by farmers in adapting to climate change in rain-fed agriculture. J Hum Ecol 44(1):23–28

11

Climate Change Resistant Energy Sources for Global Adaptation

Oluwatobi Ololade Ife-Adediran and Oluyemi Bright Aboyewa

Contents

Introduction . . . .
Vulnerability Assessment of Various Energy Systems
    Hydropower
    Solar Energy . . . . .

    Biofuel
    Oil and Natural Gas
    Thermal Power Plant
Conclusion
References

Abstract

A holistic response and adaptation to climatic vicissitudes and extreme conditions as well as their associated risks to human and ecological sustainability must adequately cater for energy needs and optimization. An interventional approach should, among other measures, seek to improve the resilience of existing and prospective energy systems to climate change. The structured and policy-driven

O. O. Ife-Adediran (✉)
Geochronology Division, CSIR-National Geophysical Research Institute (NGRI), Hyderabad, India

Department of Physics, Federal University of Technology Akure, Akure, Ondo State, Nigeria
e-mail: tobireliable@yahoo.com

O. B. Aboyewa
Department of Physics, College of Arts and Sciences, Creighton University, Omaha, NE, USA

nature of adaptation measures require a bottom-up proactive approach that envisages the performance and efficiency of these systems, especially in terms of their sensitivity and vulnerability to changing climate conditions. Therefore, this chapter seeks to scrutinize various sources of energy concerning their resistance capabilities to climate change in the face of increasing global energy demands and consumption. Renewable and conventional energy sources are co-examined and compared vis-à-vis the current trends and predictions on climatic factors that are bearing on their principles of production, supply, and distribution. Findings from this chapter will serve as assessment tools for decision makers and corroborate other ongoing discourse on climate actions towards socioeconomic development and a sustainable environment.

Keywords

Climate change · Extreme conditions · Energy resources · Sustainability · optimization

Introduction

Energy represents a fundamental requirement for wealth and job creation in many parts of the world. For many decades, humans have leveraged on the conversion of chemical energy in fossil fuels to other forms of energy. Unfortunately, these energy conversions are accompanied by the release of gases into the atmosphere and this has gradually engendered climate change; hence, the recent emphasis on energy policy alterations that factor sustainable energy supply and environment (Bang 2010). On a global scale, climate change poses increasingly severe risks on the ecosystem; the effect is particularly telling on human health, environmental safety, agriculture and economy. These changes are mostly as a result of the release of large amounts of greenhouse gases (GHG) into the atmosphere. These gases result from anthropological activities all over the world, such as burning fossil fuels for electricity and heat generation, as well as the use of internal combustion engines for transportation (EEA 2017). According to the 2019 report of the International Renewable Energy Agency (IRENA), despite the actions towards global energy transformation, carbon dioxide (CO_2) emissions that are related to the energy sector, have increased annually by over 1% on average from 2013 and 2018 (IRENA 2019). The European Environment Agency also submits that the energy sector, especially in relation to utilization of fossil fuels, lies at the heart of the challenges that are associated with climate change (EEA 2017). In fact, the utilization of energy by humans constitutes the largest contribution to the emission of greenhouse gases. The Energy Sector Management Assistance Program (ESMAP) presents the urgency of actions to control emissions and the necessity to adapt to unavoidable climate effects from the damage already induced in the biosphere due to anthropogenic GHG emissions. Energy generation and utilization do not only contribute to climate change due to greenhouse gas (GHG) emissions but are also directly influenced by its adverse effects (EUEIPDF 2017).

Environmental concerns will largely influence the generation, supply and utilization of energy in the near future; hence, the interest in alternative fuels and improvement in energy conversion technologies (Bauen 2006). In a broad sense, climate change affects energy supply and demand in terms of endowment, exploitation infrastructure and transportation. The effect of climate change on energy security ripples to other sectors such as socioeconomy, industry, and biodiversity (Ebinger and Vergara 2011). More specifically, in 2005, climate change resulted to over 10% variation in energy yield in developing countries (World Bank 2009). Changes in climatic parameters are evidential and can be observed in the lower atmospheric conditions, sea level and surface temperatures as well as topsoil wetness or dryness (Ebinger and Vergara 2011). Some predictions by the Intergovernmental Panel on Climate Change (IPCC) include an increase in global average temperature by close to 5 °C by the end of the century, fluctuating precipitation patterns and intensity, recurring extreme weather events and sea-level expansion (EUEIPDF 2017). Changes in ambient air temperature will affect the energy required for heating and cooling applications. Storms are notable for the destruction of electricity distribution infrastructure and this is being mitigated through underground cable networks. In relation to overcoming the adverse effects of climate change on clean energy generation, Bauen (2006) opined that there is great need for ingenuity, research, investment, and regulations.

Among other merits, renewable energy plays a crucial role in future low-carbon-emission plans aimed at mitigating global warming. When combined with improved energy efficiency, renewable energy is capable of achieving a 75% reduction in energy-related CO_2 emission (IRENA 2019). Bhuiyan et al. (2018) suggested that conservation of biodiversity and environmental sustainability require the exploitation of different renewable energy sources. However, the dependence of renewal energy exploration on and climatic conditions and susceptibility to climate change is a significant area of concern (Ebinger and Vergara 2011). This is in addition to their relatively high cost of generating electricity and uneven natural distribution cum availability. In principle, the adaptation of energy systems is primarily aimed at sustainable energy supply as well as balance between energy production and consumption under varying temporal and spatial conditions (IPCC 2007). However, in reality, climate variability and extremes represent the most significant threat to the entire energy supply chain. Some future projections reveal that the energy sector will become increasingly susceptible to climate change; hence, the need for effective adaptive measures (Ebinger and Vergara 2011).

Energy adaptation in the face of climate change requires reliable and sufficient weather and meteorological data, forecast models and modalities for the performance evaluation assessment of energy systems (Troccoli 2009). Climate change impacts are envisaged throughout the energy system; affecting both demand and supply. There are associated changes in demand patterns for heating and cooling due to rising temperatures. On the supply end, impacts include changes in wind direction and intensity, solar and hydropower resources, the available crops as raw materials for biomass energy production, costs and accessibility of fossil fuels in the face of frozen sea water and permafrost, the efficiency of Photovoltaic systems, thermoelectric power plants and cable systems due to increasing temperatures, as well as operational downtime due to the occurrence of extreme weather events (Cronin et al. 2018). In low income countries, energy supplies are unstable as a result of the relatively high dependence on hydroelectricity and biomass energy which are contingent on rainfall patterns and intensity (Ebinger and Vergara 2011).

Although there is an increased focus on investigating the impact of climate change on energy systems, the formal knowledge base appears to be limited (Wilbanks et al. 2007; Schaeffer et al. 2012). More importantly, studies are based on scenario analysis rather than predictions and are hindered by inherent uncertainties in modeling tools (Schaeffer et al. 2012). There are significant uncertainties that are associated with future climate projections and its effect on energy systems (Schaeffer et al. 2012). This chapter seeks to scrutinize various energy sources, especially concerning their resistance capabilities to climate change using existing research database on global and regional predictions on the vulnerability assessment of energy systems. Renewable (hydropower, wind, solar energy, biofuel etc.) and fossil energy (oil and natural gas) sources are co-examined and compared vis-à-vis the current trends and predictions on climatic factors that are bearing on their principles of production, supply and distribution.

Vulnerability Assessment of Various Energy Systems

Hydropower

Hydropower contributes significantly to total and per capita energy supply in many countries as a prominent renewable and primary energy source of electricity. The main requirement for the production of hydropower is runoff, which is largely dependent on rainfall (Hamududu and Killingtveit 2012). In view of the absolute dependence of hydropower on water resources, countries that depend largely on this source of energy experience a seasonal variation in the production of electricity (Teotónio et al. 2017; Rübbelke and Vögele 2012). Climate variability will definitely affect precipitation rate and regularity (Costa et al. 2012; Turner et al. 2017), which in turn affects hydroelectric power generation through changes in river flow, water storage, and downstream energy production (Tapiador et al. 2011; Teotónio et al. 2017). Apart from the direct effects, climate change can indirectly affect hydropower generation as a result of competitive demands and pressure on water resources by non-energy based economic sectors such as agriculture. Climate change can also engender contentions regarding water resource allocation and utilization between countries that jointly possess river catchments (Teotónio et al. 2017). Water reservoirs enhance the management of transient flow events and changes in river flow intensities (Lehner et al. 2005; Gaudard and Romerio 2014; Teotónio et al. 2017). Inter-seasonal water storages can also be adapted as resilience measures against the effect of climate change on hydropower. The effect of climate change on hydropower depends on the depths and surface areas of dams. The evaporation potential of shallow and wide dams makes them vulnerable to climate change as a result of the heat associated with global warming and the consequent increase in evaporation and drought intensity (Mukheibir 2013; Trenberth 2011; Teotónio et al. 2017).

Studies related to climate change impacts on hydropower can be done using different methods, with data that are assessed globally, regionally, or locally. Predictive impact assessments can also be carried out using modeling techniques. Although large scale analyses demonstrate inherent uncertainties based on modeling choices, they, however, provide tools for assessing global vulnerability hotspots. More impressive accuracies are achieved with models that are developed using data obtained from regions with similar climatic conditions (Turner et al. 2017). Some studies and projections found in literatures are reported as follows; Hamududu and Killingtveit (2012) used a group of simulated regional patterns of variations in runoff, which were generated from global circulation models (GCM), to estimate changes in global hydropower generation resulting from predicted changes in climate. The evaluation showed that significant impacts of climate change on hydropower generation, especially from existing power systems, is more probable on a country or regional level than a global scale. On the contrary, a recent study by van Vliet et al. (2015), using an integrated modeling and data framework on existing hydropower power plants, forecasted a decrease in usable output for more than half of the hydropower plants worldwide for a period of thirty (30) years starting from 2040. Unlike previous studies that adopted the coupled hydrological electricity modeling method on a small scale, van Vliet et al. (2015) utilized this model to establish the effect of climate change and varying water resources on global electricity supply in the twenty-first century. It is apparent that results of localized predictions of the effect of climate change on the generation of hydropower should be carefully interpreted and cannot be generalized. Furthermore, the generalization of parameters that have high spatial variation should be avoided (Schaefli 2015; Turner et al. 2017).

Turner et al. (2017) identified regions with projected significant losses in hydropower production using a coupled, global hydrological and dam operating model with three GCM projections for two emissions frameworks. The global vulnerability hotspots that were identified include; the countries of southern Europe that surround the Mediterranean Sea, as well as those within North Africa and the Middle East region. In a local study, Teotónio et al. (2017) employed an optimization model to predict the impact of climate variation on the availability of water resources and the electrical power generation in Portugal. Findings from the study showed that by 2050, hydropower generation may fall significantly and lead to higher costs of electricity in the study location as a result of climate. This result is corroborated by a similar analysis carried out by Carvajal et al. (2017) in Ecuador, which revealed that hydropower generation is relatively unstable and vulnerable to climate change since variations in throughput to hydropower stations would result in changes in the expected hydropower generation, given stable conditions.

While hydropower exploitation contributes to low carbon development initiatives, reduction in plant output due to drying climate conditions engenders the dependence on fossil fuels especially when other renewable energy sources are not competitive (Spalding-Fecher et al. 2017). The resistance of hydropower plants to climate change and its consequent effect on availability of freshwater resources can be enhanced through improvements in plant efficiency, cooling systems and fuel switches (van Vliet et al. 2015). Recently, the International Hydropower Association

(IHA), presented a six-phase approach to climate resilience for the hydropower sector which includes; comprehending what climate resilience means for the sector, a first phase of qualitative assessment of the project to climate risks, an initial or preliminary analysis, a climate stress test, risk management planning, and lastly the monitoring, evaluation and reporting of the outcomes of each of the phases (IHA 2019).

Solar Energy

Solar power ranks as the third most utilized renewable energy (after hydropower and biomass energy) and constitutes an increasingly essential component of the future low carbon-energy campaign across the globe (Panagea et al. 2014). Solar energy is mostly harnessed for both heating and electricity generation through active or passive energy systems which utilize solar collectors, photovoltaics, power towers, solar ponds, or ocean thermal collection. Through the Photoelectric effect, solar cells are utilized in the direct conversion of solar radiation to electricity; conversion efficiencies above 20% can be achieved through technologies that concentrate the incident light rays (Patt et al. 2013). The efficiency of Photovoltaic (PV) systems can be primarily influenced by geometry, ageing, ambient conditions, environmental pollution, as well as electrical, visual, and shade losses (Skoplaki and Palyvos 2009; Mani and Pillai 2010; Meral and Diner 2011; Panagea et al. 2014). The output of PV cells has a nearly linear inverse relationship with cell temperature and a directly proportional response total irradiance. Hailstorms can also affect the performance and most notably, the physical components of PV systems (the PV modules). Some of the existing solar energy technologies such as pump or circulation systems are used in environments with extreme freezing or overheating temperatures.

The impact of climate change on temperature and irradiance will significantly affect PV output (Crook et al. 2011). Climate change effects have a bearing on atmospheric transmissivity by changing atmospheric water vapor content, cloudiness (Cutforth and Judiesch 2007), and its aerosols content (Gaetani et al. 2014). Generally, the efficiency of PV modules drops by about 0.5% per degree Celsius rise in temperature (Patt et al. 2013). A study by Crook et al. (2011) utilized data obtained from coupled ocean-atmosphere climate models to indicate that PV output from 2010 to 2080 will change slightly in Algeria and Australia while it will probably rise or reduce in some parts of North America, Europe, and Asia. In another study conducted by Gaetani et al. (2014) using climate-aerosol modeling experiments, the results showed that by 2030, there will be varied effects of climate change on the production of photovoltaic energy in different parts of the world; this result is consistent with that of Crook et al. (2011). In Greece, there is an inverse relationship between the output of PV devices and the predicted rise in annual temperature, which is exceeded by the projected increase of total radiation resulting in a net increase in energy output (Panagea et al. 2014). Further location-specific studies are required for quantitative assessments of climate impacts on PV systems (Patt et al. 2013).

Wind

Wind energy is well explored for electricity generation, but it is not insulated from the influence of climatic vicissitudes since its energy density depends on global energy balance and the consequent atmospheric motion (Pryor and Barthelmie 2010; Schaeffer et al. 2012). The energy output from wind has a cubic relationship with the speed of the moving air mass, and it is proportional to air density (Manwell et al. 2002; Pryor and Barthelmie 2010). A differential change in wind speed will affect its potential for electricity production, timing, and the operational period of wind plant (Pasicko et al. 2012). According to Baker et al. (1990), a 10% change in the average wind speed could alter energy production by up to one-quarter of its initial capacity. Hence increase in wind speeds as a result of climate change should result in increased energy outputs and this could translate to an increased dependence on this energy source. Wind turbines are carefully structured to factor transient wind conditions such as the occurrence of extreme wind speeds, and other aerodynamics such as directional variations (DNV 2002). The reliability and safety of a wind power plant is determined by the maximum wind speed for which it is designed (Pasicko et al. 2012). Wind turbines should be kept nonoperational during conditions that exceed their reliability and safety indexes in other to avoid infrastructural damages.

Wind speed varies significantly with elevation from sea level as well as the spatial density of natural and artificial wind breakers (Ebinger and Vergara 2011). Several other variables can impact the vertical profile of a moving air mass, and the extrapolation of wind speeds at heights where they are not measured is quite convoluted. The logarithmic extrapolation method is popularly adopted in the estimation of wind speeds at hub heights of wind turbines that are above 50 m (Schaeffer et al. 2012). The consideration of the terrain roughness is characteristic of this method of estimation. With regards to wind energy, the roughness of a terrain is dependent on vegetative cover which can be significantly affected by climate changes and consequently impact the generation of wind power (Lucena et al. 2010; Schaeffer et al. 2012). As such, the development of wind energy production requires reliable information about the potential variations in wind energy availability as well as topographical parameters in a location of interest (Bloom et al. 2008).

Biofuel

Bioenergy (which is generated from crops are such as jatropha, sunflower, cotton, sugarcane, sorghum, and maize) is a vast and diverse energy source category since it consists of a wide range of organic fuels that can be adapted to various types of technologies to run engines, produce heat or generate electricity (Kirsten 2012). Perennials with suitable land cover are preferable as energy crops. Oilseeds are particularly used as pure plant oil and for biodiesel production while sugarcane and cereals are used in some parts of the world for the production of bioethanol as alternatives to gasoline for engines such as automobiles (Kirsten 2012), thereby portending significant contribution to the reduction of carbon footprint from transportation. The sustainability of biodiesel remains a debate as most studies come to a wide range of conclusions due to differences in approach, biomass input sources, land use and its associated change impacts, choice of system limits and functional units, as well as the methods that are adopted for allocation (Rouhany and Montgomery 2019). However, it is apparent that climate change has its toll on bioenergy production. For instance, liquid biofuels are indirectly susceptible to changes in temperature and rainfall patterns, as well as available CO_2 to crops which serve as feedstock for the production of transportation biofuels (Schaeffer et al. 2012). Average crop yields and arable land suitable for growing bioenergy crops are affected by rising temperatures and changing precipitation patterns (EUEIPDF 2017; Cronin et al. 2018). Some specific resulting adverse effects in this regard are land-area losses due to flood, salinity and dry-out influences of increased temperature (Mohammad 2013). For plants, CO_2 is an essential greenhouse gas because of its necessity for photosynthetic process. Haberl et al. (2011) explicitly examined mean cropland yield change under climate change in 2050 with and without CO_2 fertilization, and the findings indicated regional variation and increase as high as 28.22% in crop yield with full CO_2 fertilization and considerable loss (up to −16.02%) when this effect is controlled.

Oil and Natural Gas

The primary concerns regarding the utilization of fossil fuels are related to their depletion, future exhaustion, and carbon emission. Oil and gas conventional energy deposits are not likely to be affected by climate change due to their long-term formation process in geological traps (Ebinger and Vergara 2011). However, there could be indirect effects on the utilization of these resources especially with regards to the identification of natural reserves and accessibility to them (Schaeffer et al. 2012). Climate change events can affect oil and gas exploration, processing and transportation (Burkett 2011). Hurricanes and other extreme climatic events can hamper the production of these fossil fuels from off-shore facilities (Ebinger and Vergara 2011). Similarly, Burkett (2011) identifies that changes in climate variables such as sea and CO_2 levels, intensity of storms, wave regime, air and water temperature, rainfall patterns and ocean acidity also can affect oil and gas exploration in coastal regions (Schaeffer et al. 2012). There are also predictions that significant reduction in ice cover may increase the feasibility of exploration in areas of the Arctic (Harsem et al. 2011; Ebinger and Vergara 2011; Schaeffer et al. 2012), raising the prospect for oil and gas development in this region.

Thermal Power Plant

The impacts of climate change on thermal power plants are mainly related to generation cycle efficiency and water needed for cooling of power plants (Wilbanks et al. 2007; Ebinger and Vergara 2011). Some technologies that could be affected in this regard include coal, biomass residue, and geothermal power plants (Schaeffer et al. 2012). Generally, thermal plants operate with Rankine or Brayton thermodynamic cycles, require energy for heating and cooling processes, which depend on average ambient atmospheric conditions (Schaeffer et al. 2012). Increase in temperature is likely to reduce the thermal efficiency and output of power plant (Ibrahim et al. 2014; Linnerud et al. 2011; Cronin et al. 2018). Also, expected changes in water resource availability throughout the globe will have a direct influence on the use of water for cooling existing power plants and consequently, their output capacity (Wilbanks et al. 2007). Furthermore, reduction in water resources that is available for cooling may result to load-shedding or shutdown of power stations (Cronin et al. 2018). It has been projected that by 2040 capacity reductions of 12–19% in Europe and the US are possible due to rise in water temperatures and reduction in runoff (van Vliet et al. 2012; Sieber 2013). Generally, while it is expected that changes on power plants capacity will vary according to regions (for example, increase in India and Russia); generally, global annual thermal plant capacity is likely to shrink by 7–12% in the mid-century (van Vliet et al. 2015).

Conclusion

The global concern on climate change, extreme conditions and their associated footprints are in increasingly alarming degrees, especially in recent times. In spite of the preventive and adaptive measures that have been applied at local and international fronts, it is vivid that remedial actions must be taken to supplement the efforts against the adverse effects of climate change and this study shows that the energy sector is not left out of the need for intervention. The interaction between climate change and energy adaption is significant and should be taken seriously. This chapter makes a case of climate change-resistant energy utilization by examining the resilience of different energy sources to climate change, through a retrospective consideration of relevant findings from literatures on energy resources, needs and utilization as well as climatic factors, their current trends and future prediction. Consequently, it explicates the dire need for adaptive measures towards energy utilization in the face of climate changes and severe conditions. Climate change has direct and indirect effects on the exploitation of conventional and non-conventional energy resources. The effect of climate change on water resources and temperature seems to have the most significant effect on different energy sources and technologies. Extreme weather events such as hurricanes and hailstorms are detrimental to the energy systems with relatively fragile structures especially for renewable energy exploitation. This study reveals that there are temporal variations in the effect of climate change on energy systems in different parts of the world. Hence, local conditions must be considered in decision-making with regards to climate change and energy. It is hoped that this discussion provides an adequate injection of pooled examinations of different energy sources towards the sustainability and optimization of their utilization in response to climate change.

References

Mohammad Ali (2013) Climate change impacts on plant biomass growth. Springer, Dordrecht/Heidelberg/New York, pp 29–41

Baker RW, Walker SN, Wade JE (1990) Annual and seasonal variations in mean wind speed and wind turbine energy production. Sol Energy 45(5):285–289

Bang G (2010) Energy security and climate change concerns: triggers for energy policy change in the United States? Energy Policy 38:1645–1653

Bauen A (2006) Future energy sources and systems – acting on climate change and energy security. J Power Sources 157:893–901

Bhuiyan AM, Jabeen M, Zaman K, Khan A, Ahmad J, Hishan SS (2018) The impact of climate change and energy resources on biodiversity loss: evidence from a panel of selected Asian countries. Renew Energy 117:324–340

Bloom A, Kotroni V, Lagouvardos K (2008) Climate change impact of wind energy availability in the Eastern Mediterranean using the regional climate model PRECIS. Nat Hazards Earth Syst Sci 8:1249–1257

Burkett V (2011) Global climate change implications for coastal and offshore oil and gas development. Energy Policy 39:7719–7725

Carvajal PE et al (2017) Assessing uncertainty of climate change impacts on long-term hydropower generation using the CMIP5 ensemble – the case of Ecuador. Climate Change 144(4):611–624

Costa AC, Santos JA, Pinto JG (2012) Climate change scenarios for precipitation extremes in Portugal. Theor Appl Climatol 10:217–234

Cronin J, Anandarajah G, Dessens O (2018) Climate change impacts on the energy system: a review of trends and gaps. Clim Chang 151:79–93

Crook JA, Jones LA, Forstera PM, Crook R (2011) Climate change impacts on future photovoltaic and concentrated solar power energy output. Energy Environ Sci 4:3101–3109

Cutforth HW, Judiesch D (2007) Long-term changes to incoming solar energy on the Canadian prairie. Agric For Meteorol 145:167–175

DNV/Riso (2002) Guidelines for the design of wind turbines, 2nd edn. Jydsk cetraltrykkeri, Copenhagen. 286 pp

Ebinger J, Vergara W (2011) Climate impacts on energy systems: key issues for energy sector adaptation. World Bank, Washington, DC, pp 26–51

EUEIPDF – EU Energy Initiative Partnership Dialogue Facility (2017) Energy and climate change adaptation in developing countries. European Union Energy Initiative Partnership Dialogue Facility, Eschborn, pp 6–12

European Environment Agency – EEA (2017) Energy and climate change. European Environment Information and Observation Network (Eionet), Copenhagen, pp 1–10

Gaetani M, Huld T, Vignati E, Monforti-Ferrario F, Dosio A, Raes F (2014) The near future availability of photovoltaic energy in Europe and Africa in climate-aerosol modelling experiments. Renew Sust Energ Rev 38:706–771

Gaudard L, Romerio F (2014) The future of hydropower in Europe: interconnecting climate, markets and policies. Environ Sci Policy 43:5–14

Haberl H et al (2011) Global bioenergy potentials from agricultural land in 2050: sensitivity to climate change, diets and yields. Biomass Bioenergy 35(12):4753–4769

Hamududu B, Killingtveit A (2012) Assessing climate change impacts on global hydropower. Energies 5(2):305–322

Harsem O, Eide A, Heen K (2011) Factors influencing future oil and gas prospects in the Arctic. Energy Policy 39:8037–8045

Ibrahim SMA, Ibrahim MMA, Attia SI (2014) The impact of climate changes on the thermal performance of a proposed pressurized water reactor: nuclear-power plant. Int J Nucl Energy 2014:1–7

IHA – International Hydropower Association (2019) Hydropower sector climate resilience guide. http://www.hydropower.org. p 7

IPCC – Intergovernmental Panel on Climate Change (2007) Climate change 2007: synthesis report. In: Pachauri RK, Reisinger A (eds) Contribution of working groups I, II and III to the Fourth Assessment Report of the Intergovernmental Panel on Climate Change. IPCC, Geneva. 104 pp

IRENA – International Renewable Energy Agency (2019) Global energy transformation: a roadmap to 2050. The International Renewable Energy Agency, Abu Dhabi. p 9

Lehner AB, Czisch G, Vassolo S (2005) The impact of global change on the hydropower potential of Europe: a model-based analysis. Energy Policy 33:839–855

Linnerud K, Mideksa TK, Eskeland GS (2011) The impact of climate change on nuclear power supply. Energy J 32(1):149–168

Lucena AFP, Szklo AS, Schaeffer R, Dutra RM (2010) The vulnerability of wind power to climate change in Brazil. Renew Energy 35:904–912

Mani M, Pillai R (2010) Impact of dust on solar photovoltaic (PV) performance: research status, challenges and recommendations. Renew Sust Energ Rev 14(9):3124–3123

Manwell JF, McGowan JG, Rogers AL (2002) Wind energy explained: theory, design and application. Wiley, Chichester

Meral ME, Diner F (2011) A review of the factors affecting operation and efficiency of photovoltaic based electricity generation systems. Renew Sust Energ Rev 15(5):2176–2184

Mukheibir P (2013) Potential consequences of projected climate change impacts on hydroelectricity generation. Climate Change 121:67–78

Panagea IS et al (2014) Climate change impact on photovoltaic energy output: the case of Greece. Adv Meteorol 201:1–11

Pasicko R, Brankovi C, Simic Z (2012) Assessment of climate change impacts on energy generation from renewable sources in Croatia. Renew Energy 46:224–223

Patt A, Pfenninger S, Lilliestam J (2013) Vulnerability of solar energy infrastructure and output to climate change. Climate Change 121:93–10.2

Pryor SC, Barthelmie RJ (2010) Climate change impacts on wind energy: a review. Renew Sust Energ Rev 14(1):430–437

Rouhany M, Montgomery H (2019) Global biodiesel production: the state of the art and impact on climate change. In: Tabatabaei M, Aghbashlo M (eds) Biodiesel. Biofuel and Biorefinery Technologies, vol 8. Springer, Cham. https://doi.org/10.1007/978-3-030-00985-4_1

Rübbelke D, Vögele S (2012) Short-term distributional consequences of climate change impacts on the power sector: who gains and who loses? Climate Change 116:191–206

Schaeffer R, Szklo AS, Pereira de Lucena AF, Cesar M, Borba BS, Pupo Nogueira LP, Fleming FP et al (2012) Energy sector vulnerability to climate change: a review. Energy 38:1–12

Schaefli B (2015) Projecting hydropower production under future climates: a guide for decision-makers and modellers to interpret and design climate change impact assessments. Wiley Interdiscip Rev Water 2(4):271–289

Sieber J (2013) Impacts of and adaptation options to extreme weather events and climate change concerning thermal power plants. Clim Chang 121:55–66

Skoplaki E, Palyvos JA (2009) On the temperature dependence of photovoltaic module electrical performance: a review of efficiency/power correlations. Sol Energy 83:614–624

Spalding-Fecher R, Joyce B, Winkler H (2017) Climate change and hydropower in the Southern African power pool and Zambezi River basin: system-wide impacts and policy implications. Energy Policy 103:84–97

Tapiador FJ, Hou AY, de Castro M, Checa R, Cuartero F, Barros AP (2011) Precipitation estimates for hydroelectricity. Energy Environ Sci 4:443548

Teotónio C, Fortes P, Roebeling P, Rodriguez M, Robaina-Alves M (2017) Assessing the impacts of climate change on hydropower generation and the power sector in Portugal: a partial equilibrium approach. Renew Sust Energ Rev 74:788–799

Trenberth KE (2011) Changes in precipitation with climate change. Clim Res 47(1/2):123–138

Troccoli A (2009) Climate and the development community. Weather 64(1):25–26

Turner SWD, Yi J, Galelli S (2017) Examining global electricity supply vulnerability to climate change using a high-fidelity hydropower dam model. Sci Total Environ 591:663–675

Kirsten Ulsrud (2012) Bioenergy and sustainable adaptation to climate change in Africa. In: Janssen R, Rutz D (eds) Bioenergy for Sustainable Development in Africa. Springer, Dordrecht, pp 299–308

van Vliet MTH, Yearsley JR, Ludwig F et al (2012) Vulnerability of US and European electricity supply to climate change. Nat Clim Chang 2(9):676–681

van Vliet MTH, Wiberg D, Leduc S, Riahi K (2015) Power-generation system vulnerability and adaptation to changes in climate and water resources. Nat Clim Chang 6:375–380

Wilbanks TJ, Bhatt V, Bilello DE, Bull SR, Ekmann J, Horak WC, Huang YJ, Levine MD, Sale MJ, Schmalzer DK, Scott MJ, Wright SB (2007) Introduction. In: Effects of climate change on energy production and use in the United States, a report by the U.S. Climate Change Science Program and the Subcommittee on Global Change Research, Washington, DC

World Bank (2009) World development report 2009: Reshaping economic geography. The World Bank, Washington, DC

12

Building Livelihoods Resilience in the Face of Climate Change: Case Study of Small-Holder Farmers in Tanzania

Saumu Ibrahim Mwasha and Zoe Robinson

Contents

Introduction .
  Background
Description of the Study Areas and Methodology
  Study Area
  Research Methods
Results and Discussion
  Maximizing Existing Household Assets: Household Decision-Making and Female
  Empowerment
  Building the Household Assets Base
  Managing Natural and Physical Capital: Managing Slow Variables
  Livelihood Diversification
  The Role of Government in Building Small-Holder Farmer Resilience and Adaptation
  to Climate Change
Conclusion
References

Abstract

The impacts of climate change are already being felt on human and environmental systems, with the brunt of the impacts being felt by communities in the Global South, particularly small-holder farmers due to their poverty levels and greater direct dependency on natural resources for their livelihoods. Hence, there is a need to understand how to build small-holder farmers' resilience to climate change. Climate change adaptation strategies need to build livelihood resilience

S. I. Mwasha (✉) · Z. Robinson
School of Geography, Geology and the Environment, Keele University, Staffordshire, UK

in the face of climate change as well as address the factors that contribute to farmers' vulnerability. This chapter draws from a mixed-method study conducted in three villages each in a different agro-ecological zone in the Kilimanjaro region, Tanzania, to explore how to build farmers' livelihood resilience through addressing factors that increase livelihood vulnerability to climate change. These farmers' livelihoods are vulnerable because of both the impacts of climate variability on the farmers' livelihood assets and certain social and environmental structures and characteristics. Building small-holder farmers' livelihood resilience that can ensure the desired levels of livelihood outcomes in the face of climate variability and change, requires integration of strategies across household resource management as well as farm-based livelihood assets, and a holistic rather than piecemeal approach to small-holder farmers' livelihoods.

Keywords

Climate change · Adaptation · Small-holder farmer · Livelihoods · Resilience

Introduction

Background

The impacts of climate change are expected to affect human and environmental systems across the globe but the more devastating impacts are projected to occur in developing countries particularly affecting small-holder farmers (Serdeczny et al. 2017). Many small-holder farmers especially in Sub-Saharan Africa farm and live in an extremely challenging environment. The production environment is characterized by reliance on rain-fed agriculture, a low level of economic diversification, and low livelihood productivity (Output per unit of input (Yu et al. 2002)) (Di Falco and Veronesi 2013). Climate change is expected to intensify existing challenges and thus there is an urgent need for adaptation of the livelihoods of small-holder farmers to enable them to thrive in the face of climate change.

Addressing how to build resilient small-holder farmers' livelihoods through adaptation to climate change is vital for food security as well as livelihood development (Afifi et al. 2014). To address these issues, considerable emphasis is placed by researchers on describing specific locally relevant agricultural or natural resource management practices or innovations that could potentially deal with impacts of extreme events at the farm/household level. However, it is important to note that the application of these strategies is context-specific and that several constraints exist that may limit farmers' capacity to optimize their benefits. Therefore, there is a need to develop adaptation tailored to the need of that community (Ebi and Burton 2008). The context-specific adaptations result from examining the vulnerability of the target community empirically, and utilizing community experience and knowledge to examine exposure and sensitivity of the community to climate change (Ebi and Burton 2008). The context-specific adaptation strategies or practices based on examination of adaptation needs of the specific community will generate more practical measures (Ebi and Burton 2008). Adaptation based on this approach is less developed (Paavola 2008), and this research helps fill this gap by presenting potential adaptation strategies that small-holder farmers in the Kilimanjaro region of Tanzania can use to adapt to increasing climate variability. Further information about the exposure and sensitivity of these small-holder farmer communities can be found in Mwasha (2020).

Description of the Study Areas and Methodology

Study Area

The study was conducted in the Kilimanjaro region located in the North-eastern part of the Tanzania mainland. The region is divided into four agro-ecological zones; the forest reserve and Kilimanjaro mountain peak where no farming activities take place, and the highland, midland, and lowland zones where farmers are located (Fig. 1). Although all zones receive rainfall twice a year, the amount in each zone differs. Characteristics of the different zones are shown in Table 1.

Research Methods

Underpinning Theoretical Frameworks

Two lenses of analysis were used in this research; a livelihood approach drawing on the sustainable livelihood framework of the UK Department for International Development (DFID) (1999) and socio-ecological resilience drawing primarily on the work of Biggs et al. (2012, 2015). The sustainable livelihood framework is an analytical structure to facilitate understanding of broad factors that constrain or enhance livelihood opportunities, and puts people and their access to assets (financial, human, social, natural, and physical) at the center of this understanding (Reed et al. 2013). This framework also considers livelihood diversification as a risk management strategy and the role of institutions' structures and processes in shaping peoples' livelihoods (Ellis 2000). One of the main proponents of the use of a sustainable livelihood framework to studies of climate change resilience of poor communities is Tanner et al. (2015). They argued for resilience studies to incorporate a livelihood approach in order to pay attention to fundamental issues of human agency and empowerment, putting people at the center by focusing on capacities for human (rather than environmental) transformation.

Biggs et al. (2012, 2015) distinguish between resilience as a property of social-ecological systems (SES) and resilience as an approach and set of assumptions for analyzing, understanding, and managing change in SES. As a system property, they define resilience of SES as the capacity of an SES to sustain human well-being in the face of change, both through buffering shocks and also through adapting or transforming in response to change (Biggs et al. 2015). In order to analyze or build the

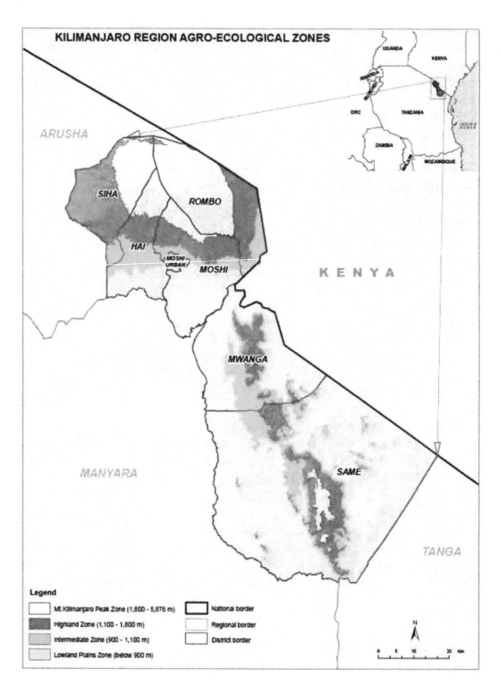

Fig. 1 Location of the Kilimanjaro region and different agro-ecological zones. (Source: Author 2020)

resilience of a system, Biggs et al. (2012, 2015) identify seven principles. The first principle refers to diversity and redundancy. Diversity involves the provision of different options for responding to change, achieved by ensuring variety (the number of different elements), balance (the number of representatives of each element), and

Table 1 Key characteristics of the three agro-ecological zones included in this study (information from Soini 2005; O'Brien et al. 2008; Regional socioeconomic profile, 2014)

Zone	Altitude (m above sea level)	Soil fertility	Rainfall (mm/a)	Temperature range (°C)	Population density (people/km^2)	Major crops	Livestock and additional comments
Highland	1100–1800	High	1250–2000	15–20	650	Wheat, beans, barley, coffee, banana, fruits, round potatoes	Majority of livestock are stall-fed. Some families own or rent plots of land in other zones particularly in the lowland zone.
Midland	900–1100	Moderate	800–1250		250	Coffee, banana, maize, beans	Dairy cattle, goats, pigs, rabbits, poultry farming
Lowland	<900		700–900	Average annual >30	50	Maize, cotton, rice, sorghum, cassava, pigeon peas	Beef cattle, goats, pigs, sheep. Provides fodder during the dry season for animals in all zones. Livestock mostly freely grazed because of the availability of open spaces especially after crops have been harvested

disparity (how different the elements are from one another). Redundancy describes the replication of elements within a system. The second principle is to manage connectivity, focusing on the way in which parts of an SES interact with each other. The third principle is to manage feedbacks and slow variables, such as long-term changes to environmental assets such as soil fertility. The fourth principle is to foster complex adaptive system thinking. The fifth principle is to encourage learning and experimentation. The sixth principle is to broaden participation and the seventh to promote polycentric governance systems. This chapter draws primarily on principles one to three, further analysis of the principles in the context of this study can be seen in Mwasha (2020).

Study Site Selection and Sampling Methods
The administrative structure in which the villages included in this study site is shown in Fig. 2. The district in which the study was conducted was selected in discussion with the regional environmental officer (who oversees activities related to the environment in the region) based on preset criteria: accessibility, evidence of climate change, and the presence of the three agro-ecological zones. The Kilimanjaro region has seven districts, but only some have three agro-ecological zones. Simple random sampling was used to select one ward from each agro-ecological zone in the selected district, and to pick one village in each ward. The district, ward, and villages are not named in order to maintain anonymity.

Data Sources, Collection Methods, and Analysis
Data were gathered from individual household surveys of household heads (who were all small-holder farmers) in the study villages, focus group discussions in each village, interviews with key informants, and researcher observations. Although the survey did not ask about the total household size because I wanted to understand the labor force, the regional socioeconomic profile (2014) shows that the Kilimanjaro

Fig. 2 Administrative structure in Tanzania

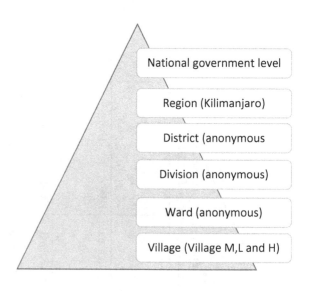

region had an average household size of 4.3 in 2012 which is the lowest in the country compared to a national average household size of 4.8 in the same year. The household surveys collected quantitative and qualitative data on the household heads' perceptions of the impact of climate change on the livelihood assets and farm production practices utilized by the household. The total numbers of interviewed households were 47, 35, and 24 in the highland, midland, and lowland zones, respectively. The total numbers of village for highland, midland, and lowland were 946, 702, and 483, respectively. Except in the lowland zone where the focus group was mixed gender, the midland and highland focus groups were gender-based. The focus group discussions explored trends in crop production, soil fertility, animal keeping, water availability, and management practices and their implication to climate change adaptation, as well as information about social factors that contribute to small-holder farmers' vulnerability to climate change. Interviews were held with five key informants (a community development officer; a representative from the Tanzania Coffee Research Institute; a district and village agriculture extension officer; and a network of farmers' groups in *Tanzania which is Nongovernmental Organization)* to understand their roles in helping farmers successfully adapt to climate change, factors that hinder farmers' adaptation ability, and potential solutions. Observations in the villages by the researcher were used to inform understanding of existing farm production practices. The data were collected in the Swahili language and translation was carried out by the researcher alongside data transcription. Details of the socioeconomic characteristics of the villages and measures contributing to livelihood vulnerability are found in Mwasha (2020) thesis.

The household survey data arises from closed-ended questions and open-ended questions with brief answers. The analysis of closed question data was carried out using SPSS descriptive statistics and qualitative data were manually coded using in vivo coding and descriptive coding, and grouped under the themes identified and entered into SPSS for analysis.

Data from the key informant interviews and focus groups were manually analyzed (Basit 2003). The transcription text was analyzed using evaluative coding based on the research questions (Smith and Firth 2011). The codes developed were categorized into different topics and then linked to form themes (Saldana 2009; Smith and Firth 2011).

Results and Discussion

This section is structured around four key areas of focus to support small-holder farmer adaptation to increase resilience to climate change. The first section focuses on the role of maximizing existing household assets, focusing specifically on human and social capital. The second section focuses on the management of physical and natural capital, also referred to as managing slow variables in reference to one of Biggs et al.'s (2015) resilience principles. The third section looks at the role of livelihood diversification, a key component of the sustainable livelihoods framework (DFID 1999), in building livelihood resilience to climate change, and Biggs et al.'s (2012, 2015) principles for building resilience. The fourth section explores the role of government in supporting small-holder farmers' adaptation and resilience to climate change.

Maximizing Existing Household Assets: Household Decision-Making and Female Empowerment

The way in which a household's assets in terms of income and produce are utilized have a major role in building livelihood resilience. Both male and female focus group participants in the highland and midland zones gave examples of how division in household responsibilities (i.e., responsibility for keeping the family fed, sheltered, educated, clothed) and divided ownership of resources within households (i.e., certain cash crops typically belonging to male members of the household) could lead to resources being used without consideration of what is best for the whole household. Focus group participants in the highland and midland zones viewed it as preferable to have joint ownership of assets and shared household responsibilities, where what is produced in the household is considered to belong to the whole family; with couples planning together how those resources are used, creating more transparent and balanced decision-making in the household. In the lowland focus group discussion, division of ownership, and household obligation were said to have no impact on household resilience. However, as the lowland focus group was mixed gender, participants may not have been willing to discuss the issues arising from different gender roles within the household. The details of division of household obligations of gender roles are found in Mwasha (2020).

Achieving more balanced decision-making around resource use within households is supported by wider policies that promote the empowerment of women through increasing access to education and financial capital. The need to support female empowerment was supported in discussions within both gender focus groups in the highland and midland zones as there was no direct question asked about this mater but it came up during the discussion about household resource use and its impact on household resilience to climate variability and change. The empowerment of women is also reported in the literature as increasing household resilience to climate change (Almario-Desoloc 2014). However, not all women in the Kilimanjaro region, and Tanzania in general, have access to these opportunities (Kato and Kratzer 2013), and the slow pace of change requires other strategies to be adopted in some households to address problems of household resource utilization stemming from household inequities. In the highland and midland, there was discussion of how some women would hide some crop produce to be able to use this in times of adversity; showing how individual household strategies are used in helping build household resilience.

Another strategy farmers in all focus groups reported to use to build resilience relates to food storage systems especially after harvest to give food a longer life and protect it from damage. The focus group participants in all three zones mentioned the use of storage tanks which are tightly sealed after being filled with food, especially maize and beans (other crops were also mentioned by female participants), as their main method of storing food, to provide reserve food stocks in times of limited harvests. Although there was no specific question asked about food storage systems, these discussions came about while exploring hunger periods in the study area.

Building the Household Assets Base

Small-holder farmer livelihoods depend on the quality of the assets (human, social, financial, and natural capitals) to which a household has access. The available assets form the foundation upon which livelihoods are built and define the ability of the people in question to execute different livelihoods strategies (Chambers and Conway 1992; DFID 1999; Ellis 2000; Scoones 2009). The discussion below explores ways to further build human and social capital (assets) which were perceived by respondents to be decreasing in the study area, and through this build resilience to climate change.

Building Human Capital

Human capital can be defined as the available labor force within a household to contribute to agricultural production, income generation, and household management. The household surveys identified that human capital was negatively affected as a result of malaria which has intensified by increasing temperatures allowing the geographic spread of mosquitoes particularly into the highland zone.

Adaptation strategies are needed to deal with an increase in malaria and its impact on human capital (Onwujekwe et al. 2000; Teklehaimanot and Mejia 2008; Asenso-Okyere et al. 2011). Teklehaimanot and Mejia (2008) summarized malaria control strategies as follows:

- Provision of early diagnosis and prompt treatment
- Selective and sustainable use of preventive measures, including vector control
- Prevention, early detection, and containment of epidemics
- Strengthening local abilities and applied research

These strategies require actions at both government and household level. Households are responsible for using preventive measures such as mosquito bed nets, managing the environment to reduce mosquito habitats, and going to the hospital when they get ill for diagnosis and treatment. The government is responsible for educating people about control measures, ensuring access to control measures and medical services, and investing in research on prevention and treatment of malaria diseases. Poverty also plays a crucial role in explaining why certain population groups are more vulnerable to malaria, because of the inability to pay for insecticide-treated bed nets and access to medical health (Teklehaimanot and Mejia 2008). In the study area, the household survey showed that most families were provided with free mosquito nets by the government to prevent malaria disease.

Building Social Capital

The household surveys in this study show a perceived decrease in social capital (described as support that households provide to each other) partly caused by climate variability reducing the amount of crop yield. Social capital is important as it can help households survive the impacts of poverty (Baiyegunhi 2014) through providing security in times of distress and access to resources (Grech 2012). There are different ways that households or individuals can build social capital. Baron et al. (2000) identified two different aspects of social relations that can build social capital for households and individuals: (i) obligations, expectations, and trustworthiness; and (ii) norms and effective sanctions.

One core element of social capital is where people are willing to help each other and do things for each other, relating to Baron et al.'s (2000) "obligations, expectations, and trust." For example, if an individual (named A) does something for B, and trusts B to reciprocate in the future, this establishes expectations in A and an obligation on the part of B. As A does the same to more people, and these people are trustworthy and responsible, this creates good safety nets for A in the event that something happens to A. This study revealed one practice based on these concepts of social capital, in the saving of food resources by one household for another, to be taken back in times of need. This area of social capital was the preserve of females in the communities who reported to support each other in this way.

Social capital can also be developed through the development of norms and sanctions (Baron et al. 2000). However, these can be fragile as some people can misuse them for personal interest. Norms and sanctions relevant to this study include the promotion of norms that encourage family members and neighbors to act selflessly in support of others, which may be reciprocated, providing positive motivations. For example, a household which does not support another household may not be supported if they need help themselves (negative motivation), while the household which supports another during difficult times may be positively motivated by the possibility of needing support themselves in the future.

These aspects that develop social capital for households and individuals are strengthened by social structures with high interconnectedness and interdependency between all actors (Fig. 3; Baron et al. 2000). This can be seen as another form of social capital; the connectivity of social structures that allow the proliferation of obligations and expectations. For example, in a less connected "open structure"

Fig. 3 Social networks without closure (**a**) and with closure within the social networks (**b**). (Source: Baron et al. 2000)

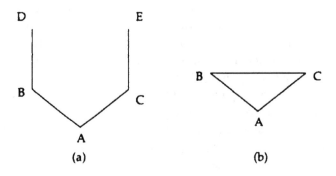

(Fig. 3a) individual A can carry out actions that negatively affect D and E who are not acquainted with each other, and therefore cannot unite to control the negative effects from A. In contrast, Fig. 3b shows that all actors are connected and therefore there will be stronger motivation for A to avoid negative effects on other actors.

The household survey and information from the community development officer shows that in the study area some households have developed structures that facilitate social capital. They support each other by forming groups based on similarities between members such as family relations, friends, or similar work space or type. The groups work as informal saving and credit institutions, and members get different services such as access to loans and social support in case they face shocks in their life. Based on information from key informants, there is some nongovernment organization capacity building of these groups by providing training into how groups select leaders and develop financial management and group policy. The government encourages group members to register through the community development office to increase accountability but most are reluctant to register because of registration fees. More government-subsidized training and waiving of registration fees could improve social capital in the area.

Drawing on the insights from Baron et al. (2000), households and individuals can improve their social capital by their own acts to support others and by developing trusted networks of individuals and households within their community, increasing their sources of support for times of adversity. Village-level initiatives based around these principles could also help develop social capital through promoting norms, trust, and responsibility and interconnectedness at the community level. Government-level initiatives can also help support more formalized social support networks such as credit unions.

Managing Natural and Physical Capital: Managing Slow Variables

This subsection adopts the concept of managing "slow variables" as used in Biggs et al.'s (2012; 2015) resilience framework. Slow variables are those variables that change slowly and can take time for changes in them to be noticed. The slow variables considered in this study focus on soil and water, two key areas of natural capital on which small-holder farmers depend, and which the results show have been affected negatively by management practices. Drawing on the study's focus group discussions, household surveys, key informant interviews, and wider literature, this section describes strategies to ensure the adequate supply of water for agriculture, and strategies to improve soil fertility in the face of declines in these "slow variable" assets.

Managing Water Resources

Poor decisions and practices made by farmers, such as crop choice and inadequate water source management practices, can affect agricultural productivity particularly under weather-related stress (Bot and Benites 2005). This section looks at how farmers will need to adapt to climate change by choosing appropriate crops to grow by considering plant traits with tolerance of water stress, conservation of water sources, and employing farming practices that encourage water infiltration and conserve soil moisture.

Planting Crops That Can Survive Climate Variability

Data collected in the study area demonstrated that in all three zones the types of agricultural produce grown by farmers threatens the capacity of the existing water resources to sustain agricultural production in the area, particularly as available water resources are perceived to have decreased. Cultivation of high value, high water demand crops, particularly horticultural produce such as tomatoes, cucumbers, onions, and carrots, has increased, increasing pressure on water resources. Although there are benefits to practicing horticulture as produce can be harvested in a relatively short time, the viability of these choices of crop is in question because of the capacity to sustain production in the long-term with trends of declining water availability. In contrast, traditional maize varieties are perceived to take a long time to mature (~6 months) which makes them vulnerable to the more variable rainfall patterns, potentially preventing the crops from reaching maturity. One solution is to replace traditional maize seeds with early maturing maize varieties which can survive the increased rainfall variability. In addition, planting drought-resistant crops such as millet is another strategy for managing the effects of decreasing rainfall amount and increasing variability. However, the results from this study show that the use of these crops is low in all three zones. As some farmers have negative attitudes toward crops like millet and cassava (another drought-resistant crop), there is the need for the government to encourage farmers to grow these crops through the development of an effective market for these crops. Government can also play a role in supporting further research into alternative drought-resistant crops but with a need to ensure that these will meet the needs and wants of small-holder farmers.

Conservation of Water Sources

Across all three zones, there are a number of different practices that were discussed as being important for the conservation of water resources, including encouraging on-farm tree planting and avoiding the cutting down of trees. The relationship between trees and water resources is complicated (Ellison et al. 2017). However, there is some indication that increasing tree planting on farms can help preserve water resources through the effect of shading to reduce evaporative loss from soils (Clement et al. 2016).

The use of farming techniques for increasing infiltration, reducing surface runoff and evaporation, and improving soil water availability is essential in dealing with weather-related shocks (Biazin et al. 2012). There was evidence of these practices in the study area. For example, in the highland zone the use of terraces was reported to help control soil erosion, as well as reducing surface runoff and increasing infiltration (Biazin et al. 2012). In the lowland zone, farmers reported using minimal tillage in order to protect soil moisture.

Rainwater Harvesting

Rainwater harvesting may also increase productivity from existing rainfall. The harvesting systems mentioned by focus group participants can be categorized into two types: earth dams and spate irrigation. Focus group participants in all three zones reported the need for earth dams especially in the lowland zone, to harvest available rainwater and use it for supplemental irrigation.

The use of spate irrigation was mentioned in the focus group in the lowland zone, where floodwaters from the highlands are channeled to nearby fields through gravity or water pumps to irrigate farms in the lowlands. This approach is used to cope with dry spells by taking advantage of the rain happening in upland areas even though it has ceased in lowland areas, as the rainy seasons last longer in the highland and midland zones relative to the lowland zone.

In addition to providing additional water sources, farmers with access to spate irrigation reported other benefits such as improved soil fertility, reducing the use of fertilizer on their farms because the spate water comes with eroded materials that are nutrient-rich. Although this strategy has many benefits it may come with other challenges because large flash floods can potentially cause damage to crops and prepared land (Komakech et al. 2012), although this was not mentioned by any focus group or household survey participant.

In addition to direct interventions such as water-conservation farming practices, wider systemic issues leading to poverty need to be addressed, because in some cases it is these systemic issues that push farmers in this area to inappropriately use water sources as a survival strategy. For example, as a result of reliable crop market, farmers chop down trees to sell timber as alternative source of income.

Managing Soil Fertility

Management of soil fertility particularly where weather-related shocks are experienced is vital for reducing the impact of climate change on soils as well as improving water productivity (Biazin et al. 2012). Soil fertility-related land management practices include appropriate use of fertilizer and soil conservation methods to reduce erosion and maintain organic matter (Clair and Lynch 2010; Kaczan et al. 2013). Farmers in the study area identified strategies to increase soil fertility and conserve soil moisture such as mulching by retaining crop residues in fields, intercropping, the use of organic fertilizers, minimal tillage, and agroforestry as potential adaptation strategies. However, comparison between zones shows that while retaining crop residues, intercropping and use of organic fertilizers were mentioned in all three zones, minimal tillage was not discussed as a strategy in the midland and highland zones. In addition, the role of agroforestry was debated in the lowland zone, with the majority of participants believing agroforestry to be appropriate in the midland and highland zones but not in the lowland zone because they perceive the crops grown in the lowland zone (see Table 1) do not require the shade provided by trees.

Focus group participants believed that given the increased nature of climate variability and projected climate change, the benefits from retaining crop residues in fields were more important than ever. However, although maintaining crop residues in fields was seen as an important potential adaptation measure to build livelihood resilience to climate variability, it was also acknowledged that this is hindered by the free grazing of animals by livestock keepers residing in the lowland zone. This was particularly problematic for farmers in the highland zone who also held farmland in the lowland zone, and were not able to monitor their land in the lowland zone. In addition, the need for fodder for livestock kept in the midland and highland zones was partly met by the transport of crop residues from the lowland zone. Participants reported that the nature of the issues faced meant that farmers now needed to find the balance between livestock feed and improving agricultural soils through retaining crops residue.

In dealing with competing crop residue demands, agricultural intensification can help increase the amount of biomass produced which can be divided between livestock and that which can be retained in the soils. In addition, free-grazing animals in the lowland zone should be discouraged to motivate farmers to retain part of the residues in the farms. Livestock keepers should be encouraged to sell some of their livestock and retain only that which they can maintain using their own resources.

Agroforestry was also mentioned as a potential adaptation strategy to address issues of declining soil fertility in the study area in all focus group discussions. The household survey results showed that participants believed this strategy to have both socioeconomic and environmental benefits. Research literature has also reported several benefits of agroforestry, including an increase in soil organic matter, erosion control, reduced sensitivity to harsh weather, natural pest and disease control, and provision of an alternative source of income (Reyes et al. 2005; Nair 2007; Nguyen et al. 2013; Pumariño et al. 2015; Sepúlveda and Carrillo 2015; Schwab et al. 2015). It is important to acknowledge that, in the survey results, there were a small number of respondents who disagreed about the benefits of agroforestry (particularly in the lowland zone) when combined with cereal crops like maize. However, studies do suggest that there is potential for agroforestry within maize production (e.g., Garrity et al. 2010), which offers opportunity to explore further types of trees that could be relevant in the study area to be integrated with maize, but the acceptance of farmers would need to be developed.

The use of organic fertilizer from livestock is another potential strategy to enhance agricultural productivity (Clair and Lynch 2010) and was reported by focus group participants and key informants. However, farmers reported that the main challenge of using organic fertilizer was the inconvenience for some associated with transferring manure from the homestead where cattle are kept to the farm fields which could be up to 5 km from households. There are several potential ways to address these issues, including farmers cooperating with neighboring farm owners to hire transport and share the transportation costs. This shows the importance of social capital (in terms of a strong supportive community willing to work collectively) to address a range of issues. However, it should be acknowledged that any additional costs may be prohibitive for the poorest farmers highlighting that mechanisms to address financial capital and systemic issues leading to poverty underpin many different solutions.

Livelihood Diversification

Small-holder farmers in the study area have access to three different areas of livelihood contribution: (i) crop production for subsistence, (ii) livestock keeping for subsistence, and (iii) off-farm income activities, including small business such as small shops, street vending, and sale of agricultural products for income. The results from this study show that not all households diversify their livelihoods even though in the focus group discussions in all three agro-ecological zones it was reported that households should ensure they have more than one livelihood option as a risk management strategy, particularly important in the face of increased climate change variability. Given the types of livelihood options practiced in the study area, there are some similar dependencies in almost all livelihood options. For example, crop production and livestock keeping are all dependent on natural capital such as soil fertility and water resource availability. This suggests that if the flow and stock of these resources are affected by climate change or climate variability, the main livelihood contributions will be affected even where there is some diversity in income. Drawing on Biggs et al.'s (2012, 2015) resilience principles, it is clear that greater resilience can be achieved with greater disparity in the diversification options, and therefore having different livelihood options based around agriculture may provide less resilience than including diversification away from agriculture.

The Role of Government in Building Small-Holder Farmer Resilience and Adaptation to Climate Change

This subsection looks at the role of government in helping farmers to adapt to climate change in the Kilimanjaro region of Tanzania. The discussion is divided into two areas: (i) the role of government officials (e.g., agriculture extension officers and community development officers); (ii) government policies and directives. Nongovernmental Organizations also often play an important role in tackling livelihood adaptations in the Global South (e.g., Kajimbwa 2006; Oshewolo 2011). However, the role of NGOs was not a focus of this study and only one NGO was included as a key informant interview, nor were NGOs brought up by study participants in the discussions of livelihood resilience and adaptation to climate change. Government officials provide small-holder farmers with access to different services. For example, connections between farmers and agriculture extension services are an important way of providing farmers with timely and relevant knowledge on agricultural aspects of livelihood management (Fosu-Mensah et al. 2012) especially where traditional knowledge systems do not work in responding to climate change (Shackleton et al. 2015). The need to improve access to agriculture extension officers was mentioned by some respondents in the household survey as well as in key informant interviews. Access to extension officers by small-holder farmers could be improved by allocating more extension service providers to farmers as well as providing the service providers with improved means of transport to facilitate their transport to the small-holder farmer villages. The community development officers differ to agriculture extension officers in that they work with farmers to provide financial capital. However, the existing financial support given to women and youth group projects was reported to be inadequate compared to the demand. Therefore, the government needs to provide other sources of support such as credit, which farmers can exploit as access to financial capital is essential to facilitate adaptation (Deressa et al. 2011; Ndamani and Watanabe 2015; Belay et al. 2017).

There are a number of policy areas that were identified in the study where government could make adjustments to create more favorable conditions to support farmers' adaptation to climate change. These include: the 2007 Warehouse Receipt System (WRS) introduced to help farmers take advantage of price fluctuations by enabling farmers to store crops in warehouses and sell them when prices are high (MAFAP 2013). There was low use of WRS, especially for cereal crops. The government needs to increase the capacity of cooperatives unions, farmers' organizations, and savings and credit cooperatives to be able to implement the WRS for the variety of crops that are produced in the study area.

There are several areas of regulations where there is weak implementation causing availability of fake agricultural inputs in the market. Examples include, the Seeds Act (2003) which regulates the production and trade of all varieties of agricultural seeds including the mandatory provision of seeds for quality assurance; the Fertilizer Act (2009) regulates and controls the quality of fertilizer, either domestically produced or imported; and the Tropical Pesticides Research Institute Act (1979) which regulates research on pesticides for the purpose of ensuring their quality. The study participants reported poor implementation of these regulations with negative effects on the farmers. This shows a need for the government to ensure existing laws and policies are implemented more effectively to achieve the intended results. There were also government actions which small-holder farmers directly referred to as having a negative effect on their livelihoods. These included an export ban on crops preventing farmers from maximizing income in times of food scarcity in other countries (but supporting food security within Tanzania); government appropriation of resources (particularly water) for urban areas; and the lack of implementation of a requirement for using scales for measurement of agricultural produce meaning that farmers often ended up being short-changed by unscrupulous buyers.

Conclusion

This chapter discusses areas of focus to build the livelihood resilience to climate change of small-holder farmers in the Kilimanjaro region of Tanzania, drawing on results from a study of three villages in the region and key informant interviews. Building small-holder farmers' livelihood resilience needs to be tackled at different spatial and temporal scales; from individual farmer decisions to government interventions, from short-term to long-term strategies, and in different spheres from household management practices, agricultural practices, livelihood diversification, to the building of community and social capital, and the development of wider regulatory and economic systems.

These results highlight the importance of scale and context. Even a single region, such as the Kilimanjaro region studied here, is not a homogenous block. Although there were some similarities across the three agro-ecological zones such as the importance of measures to build social capital, control malaria disease, and the overall management of slow variables, there were some distinctions in the actual practices for management of slow variables as well as the existing social structures affecting resilience within the household. For example, the use of terraces and agroforestry was perceived as relevant in the highland and midland zones while leaving crop residues, minimal tillage, and drought-resistant crops were articulated more in the lowland zone. However, it should also be remembered that these zones do not exist in isolation, for example, some small-holder farmers own land in a different zone to the one in which they live, water resources from the highland zone travel through the midland and lowland zone, and forage for livestock may be sourced from a different zone. Therefore, changes implemented in one zone also have the potential to impact other zones, requiring adaptation measures to both respond to the specific context of an area and understand the interconnections between areas.

Combining a socioecological resilience framework and sustainable livelihood framework provides an important approach to understand small-holder farmers' assets bases as well as their vulnerability and sensitivity to climate change (Mwasha 2020). This combined framework also helps to understand the barriers to climate change adaptation strategies as well as strategies to build resilience to climate change across different temporal and spatial scales and spheres. Building resilience principles into sustainable livelihoods thinking clearly has a role to play in addressing the resilience to climate change of small-holder farmer livelihoods in the study region and likely throughout the Global South.

References

Afifi T, Liwenga E, Kwezi L (2014) Rainfall-induced crop failure, food insecurity and out-migration in same-Kilimanjaro, Tanzania. Clim Dev 6(1):53–60. https://doi.org/10.1080/17565529.2013.826128

Almario-Desoloc MS (2014) Women empowerment and climate change adaptation in northern Quezon. Asia Pac J Multidiscip Res 2(4):14–21

Asenso-Okyere K, Asante FA, Tarekegn J, Andam KS (2011) A review of the economic impact of malaria in agricultural development. Agricultural economics 42(3):293–304

Baiyegunhi LJS (2014) Social capital effects on rural household poverty in Msinga, KwaZulu-Natal, South Africa. Agrekon. Taylor & Francis 53(2):47–64

Baron S, Field J, Schuller T (2000) Social capital: critical perspectives. Oxford University Press, Oxford

Basit T (2003) Manual or electronic? The role of coding in qualitative data analysis. Educ Res. Taylor & Francis 45(2):143–154

Belay A et al. (2017) Smallholder farmers adaptation to climate change and determinants of their adaptation decisions in the Central Rift Valley of Ethiopia. Agriculture and Food Security. BioMed Central 6(1):1–13. https://doi.org/10.1186/s40066-017-0100-1

Biazin B, et al. (2012) Rainwater harvesting and management in rainfed agricultural systems in sub-Saharan Africa – a review. Phys Chem Earth. Elsevier Ltd, 47–48:139–151. https://doi.org/10.1016/j.pce.2011.08.015

Biggs R et al (2012) Toward principles for enhancing the resilience of ecosystem services. Annu Rev Environ Resour 37:421–448

Biggs R, Schlüter M, Schoon ML (2015) Principles for building resilience: sustaining ecosystem services in social-ecological systems. Cambridge University Press, Cambridge, MA

Bot A, Benites J (2005) The importance of soil organic matter: key to drought-resistant soil and sustained food production. Food & Agriculture Organization of the United Nations, Rome

Chambers R, Conway G (1992) Sustainable rural livelihoods: practical concepts for the 21st century. Institute of Development Studies (UK)

Clair SBS, Lynch JP (2010) The opening of Pandora's Box: climate change impacts on soil fertility and crop nutrition in developing countries. Plant and Soil 335(1–2):101–115

Clement M, Mwakalila S, Norbert J (2016) Implications of land use and climate change on water balance components in the Sigi Catchment, Tanzania. J Geograph Assoc Tanzania 39(1)

Deressa TT, Hassan RM, Ringler C (2011) Perception of and adaptation to climate change by farmers in the Nile basin of Ethiopia. J Agric Sci. Cambridge University Press 149(1):23–31

DfID UK (1999) Sustainable livelihoods guidance sheets. London: DFID, 445

Di Falco S, Veronesi M (2013) How can African agriculture adapt to climate change? A counter-factual analysis from Ethiopia. Land Econ 89(4):743–766. https://doi.org/10.3368/le.89.4.743

Ebi KL, Burton I (2008) Identifying practical adaptation options: an approach to address climate change-related health risks. Environ Sci Policy. Elsevier 11(4):359–369

Ellis F (2000) Rural livelihoods and diversity in developing countries. Oxford university press, Oxford

Ellison D, et al. (2017) Trees, forests and water: cool insights for a hot world. Glob Environ Change. Elsevier 43:51–61

Fosu-Mensah BY, Vlek PLG, MacCarthy DS (2012) 'Farmers' perception and adaptation to climate change: a case study of Sekyedumase district in Ghana. Environ Dev Sustain. Springer 14(4):495–505

Garrity DP, et al. (2010) Evergreen agriculture: a robust approach to sustainable food security in Africa. Food Secur. Springer 2(3):197–214

Grech S (2012) Disability, communities of poverty and the global south: debating through social capital. In: Inclusive communities. Brill Sense, Boston, pp 69–84

Kaczan D, Arslan A, Lipper L (2013) Climate-smart agriculture: A review of current practice of agroforestry and conservation agriculture in Malawi and Zambia. Rome: United Nations Food and Agriculture Organization

Kajimbwa M (2006) NGOs and their role in the global south. Int J Not-for-Profit L 9:58

Kato MP, Kratzer J (2013) Empowering women through microfinance: evidence from Tanzania. ACRN J Entrepreneurship Perspect 2(1):31–59

Komakech HC, Van der Zaag P, Van Koppen B (2012) The Last Will Be First: Water Transfers from Agriculture to Cities in the Pangani River Basin, Tanzania. Water Alternatives 5(3)

MAFAP (2013) Review of food and agricultural policies in the United Republic of Tanzania. MAFAP Country Report Series, FAO, Rome, Italy

Mwasha SI (2020) Livelihoods, vulnerability, and adaptation to climate change of small-holder farmers in Kilimanjaro, Tanzania. Unpublished PhD thesis, University of Keele, UK

Nair PR (2007) The coming of age of agroforestry. Journal of the Science of Food and Agriculture 87(9):1613–1619

Ndaman F, Watanabe T (2015) Farmers' perceptions about adaptation practices to climate change and barriers to adaptation: A micro-level study in Ghana. Water 7(9):4593–4604

Nguyen Q, Hoang MH, Öborn I, van Noordwijk M (2013) Multipurpose agroforestry as a climate change resiliency option for farmers: an example of local adaptation in Vietnam. Climatic change 117(1–2):241–257

O'Brien G, et al. (2008) Climate adaptation from a poverty perspective. Clim Policy. Taylor & Francis 8(2):194–201

Onwujekwe O, Chima R, Okonkwo P (2000) Economic burden of malaria illness on households versus that of all other illness episodes: a study in five malaria holoendemic Nigerian communities. Health policy 54(2):143–159

Oshewolo S (2011) Limited policy engagement of non-governmental organizations: a guilt trip on the Nigerian state. J Sustain Dev Africa 13(1):142–152

Paavola J (2008) Livelihoods, vulnerability and adaptation to climate change in Morogoro, Tanzania. Environ Sci Policy. Elsevier 11(7):642–654

Pumariño L, Sileshi GW, Gripenberg S, Kaartinen R, Barrios E, Muchane MN, Midega C, Jonsson M (2015) Effects of agroforestry on pest, disease and weed control: a meta-analysis. Basic and Applied Ecology 16(7):573–582

Reed MS, et al. (2013) Combining analytical frameworks to assess livelihood vulnerability to climate change and analyse adaptation options. Ecol Econ. Elsevier 94:66–77

Reyes T, Quiroz R, Msikula S (2005) Socio-economic comparison between traditional and improved cultivation methods in agroforestry systems, East Usambara Mountains, Tanzania. Environ Manag. Springer, 36(5):682–690

Saldana J (2009) Chapter 1: An introduction to codes and coding. In J Saldaña (Ed.) The coding manual for qualitative researchers (pp. 3–21). Thousand Oaks, CA: Sage

Schwab N, Schickhoff U, Fischer E (2015) Transition to agroforestry significantly improves soil quality: A case study in the central mid-hills of Nepal. Agriculture Ecosystems & Environment 205:57–69

Scoones I (2009) Livelihoods perspectives and rural development. J Peasant Stud. Taylor & Francis 36(1):171–196

Sepúlveda RB, Carrillo AA (2015) Soil erosion and erosion thresholds in an agroforestry system of coffee (Coffea arabica) and mixed shade trees (Inga spp and Musa spp) in Northern Nicaragua. Agriculture Ecosystems & Environment 210:25–35

Serdeczny O et al. (2017) Climate change impacts in Sub-Saharan Africa: from physical changes to their social repercussions. Regional Environmental Change 17(6):1585–1600. https://doi.org/10.1007/s10113-015-0910-2.

Shackleton S, et al. (2015) Why is socially-just climate change adaptation in sub-Saharan Africa so challenging? A review of barriers identified from empirical cases. Wiley Interdiscip Rev Clim Chang. Wiley Online Library, 6(3), pp. 321–344

Smith J, Firth J (2011) Qualitative data analysis: the framework approach. Nurse researcher. RCN Publishing Company 18(2):52–62

Soini E (2005) Land use change patterns and livelihood dynamics on the slopes of Mt. Kilimanjaro, Tanzania. Agric Syst. Elsevier 85(3):306–323

Tanner T, et al. (2015) Livelihood resilience in the face of climate change. Nat Clim Change. Nature Publishing Group 5(1):23

Teklehaimanot A, Mejia P (2008) Malaria and poverty. A N Y Acad Sci. Wiley/Blackwell (10.1111), 1136(1):32–37

Yu B, Fulginiti LE, Perrin RK (2002) Agriculture productivity in Sub-Saharan Africa. Annual Meetings of the American Agricultural Economics Association. California

13

Intangible and Indirect Costs of Adaptation to Climate Variability Among Maize Farmers: Chirumanzu District, Zimbabwe

Dumisani Shoko Kori, Joseph Francis and Jethro Zuwarimwe

Contents

Introduction: Adaptation, an Overview
The Cost Associated with Adaptation to Climate Variability
Location and Climate of Chirumanzu Resettlement Areas and Rationale for Selection
Characteristics of A1 Maize Farmers in Chirumanzu Resettlement Areas
Data Gathering
Data Handling and Analysis
Adaptation Measures Adopted by Maize Farmers in Resettlement Areas of Chirumanzu

    Use of Drought Tolerant Varieties
    Good Crop Establishment Practices (No Regret Measures)
    Conservation Farming
    Wetland Farming
    Diversification
Problems and Dangers Associated with Adaptation Measures Adopted by Maize

    Threat to Safety and Wellbeing
    Lack of Mechanical Equipment
    High Cost and Scarcity of Inputs
    Increased Risk and Uncertainty
    Labor Related and Timing Issues
    Negative Effects on Productivity and Profi

D. S. Kori (✉) · J. Francis · J. Zuwarimwe
Institute for Rural Development, University of Venda, Thohoyandou, South Africa
e-mail: joseph.francis@univen.ac.za

Intangible Costs Associated with Adaptation to Climate Variability
    Health, Wellbeing, and Safety Concerns
    Burden on Household Members
    Worry, Anxiety, and Uncertainty
    Pain and Suffering
    Effort Spent to Facilitate Smooth Implementation of Adaptation Measures
    Indecision During Allocation of Scarce Resources
    Ridicule and Embarrassment Associated with Borrowing Equipmen
Indirect Costs of Adaptation to Climate Variability
    Family Labor Opportunity Cost
    Additional and Unplanned Cost of Production ...
    Yield Losses Related to Risk, Unavoidable Delays, and Timing
    Huge Investment Cost for Some Adaptation Practice
Conclusion
References

Abstract

Maize farming in resettlement areas of Chirumanzu District of Zimbabwe is vulnerable to climatic variations. The Government of Zimbabwe encourages maize farmers in resettlement areas to adapt to climate variability through conservation farming and diversification among other measures. It is envisaged that the measures will improve maize farmers' resilience and ability to safeguard food and nutrition security in the country. However, the process of adaptation is dynamic, complex, and multifaceted in nature. Several problems and dangers accompany the process of adaptation. The problems and dangers are associated with intangible and indirect costs. The focus of this chapter is to explore intangible and indirect costs associated with measures adopted by maize farmers in resettlement areas of Chirumanzu in Zimbabwe. Fifty-four maize farmers from four resettlement wards provided the data through semi-structured interviews. Diversification, changing planting dates, use of drought tolerant varieties were some of the measures adopted. Several problems and dangers accompanied the adaptation measures adopted. Intangible costs such as pain and suffering, embarrassment, ridicule, and stereotyping were experienced. Indirect costs including additional and unplanned costs were also encountered. This chapter concludes that intangible and indirect costs associated with adaptation may result in reduced adaptive capacity and resilience of maize farmers. Therefore, national governments should exercise extreme caution and desist from only encouraging farmers to adapt. Rather, they should consider intangible and indirect costs involved while providing solutions to reduce them to avoid situations where farmers are worse off while facilitating sustainable adaptation.

Keywords

Unintended adaptation effects · Nonmarket adaptation costs · Smallholder farming community · Maize farming · Resettlement areas

Introduction: Adaptation, an Overview

Adaptation is an appropriate way to build resilience to climate variability (Biagini et al. 2014; Costinot et al. 2016; Menike and Arachchi 2016) especially under smallholder maize farming. As such, adaptation has been broadly accepted as a policy priority, which explains why it has received extraordinary attention from politicians (Basset and Fogeliman 2013). Following the refusal of some well-developed nations to support the greenhouse gas emission goals of the Kyoto Protocol of 2001, adaptation emerged as the major viable option for furthering the designing of the Climate Change Policy (CCP) (Schipper 2009). The Intergovernmental Panel for Climate Change (IPCC) (2014) states that it recognizes adjustments made even by smallholder farmers in an attempt to reduce vulnerability of farming activities.

Following the agrarian land reforms introduced in 1980 and the Fast Track Land Reform Programme (FTLRP) of 2000 in Zimbabwe, a vulnerable community of farmers emerged in the professed resettlement areas. Since then, the country has been experiencing food insecurity challenges, with almost 50% of the population being vulnerable to hunger due to the combined effect of unsustainable land reforms and extreme climate variations (Sachikonye 2003). As such, concerns about food and nutrition security have seen the Government of Zimbabwe (GoZ) put more emphasis on adaptation. The Zimbabwe National Climate Change Response Strategy of 2015 acknowledges that farmers in resettlement areas are vulnerable to climatic variations that are currently prevailing in the country. Thus, the GoZ is encouraging maize farmers in resettlement areas to adapt to climate variability so that they can improve resilience and ability to fulfill the massive role of safeguarding food and nutrition security.

Major success stories on adaptation to climate variability have been documented around the globe including the African region and in Zimbabwe. Overall, adaptation has improved maize yields by an average of 15–18%, although effectiveness of measures varies significantly across regions (IPCC 2014). Rurinda et al. (2013) showed that improved timing of planting and adjusting soil nutrient inputs stabilize maize yields under variable rainfall conditions in Zimbabwe. However, it is important to note that adaptation is an investment (Adam and Wiredu 2015) associated with costs as the process introduces new ways of doing things thereby calling for some tradeoffs between new and old ways.

The Cost Associated with Adaptation to Climate Variability

The process of adaptation is dynamic, complex, and multifaceted in nature. This is because it occurs in biophysical, technical, social, and psychological dimensions that are not static but evolving. Adaptation initiatives are associated with costs (Arfunuzzaman et al. 2016). The IPCC (2014) confirms and highlights that it is costly to adapt to climate variability especially when resources are scarce and capacity is limited which is the case with most of the maize farmers in resettlement areas of Chirumanzu District of Zimbabwe. Literature on climate adaptation, in general, does not fully acknowledge intangible and indirect cost of implementing adaptation plans (Milman and Arsano 2014). Yet failure to include the intangible and indirect costs would result in underestimates and misrepresentations of the total cost of adaptation. Adaptation cost literature is still evolving (Fankhauser 2009; Agrawala et al. 2011; Doczi and Ross 2014). It is regarded scant, uncertain, and consensus on overarching cost estimates is lacking (Kumar et al. 2010). There is little peer-reviewed literature on the subject.

Attempts have been made to estimate total adaptation cost. However, adaptation cost estimates only exist at a global level. It is estimated that total adaptation cost ranges from USD9 billion to USD109 billion per year by 2030 (Agrawala et al. 2008; Chambwera et al. 2014). For the agricultural sector, adaptation cost estimates are rare. Agrawala et al. (2011) indicated that with the exception of Mccarl (2007) literature on the cost of adaptation in agriculture is lacking. Mccarl (2007) used a top-down approach to estimate the cost of adaptation in the Agriculture, Forestry, and Fisheries sector. It is estimated that adaptation will cost USD14.23 billion per year by 2030. It is clear that methods that have been used to estimate adaptation costs are quantitative giving much attention to attaching a monetary value to tangible and direct costs. Intangible and indirect costs that do not have a market value are rarely considered.

Agrawala et al. (2008) also observed that there are no accepted metrics for assessing the cost of adaptation measures. As such, multiple definitions for adaptation cost exist. Existing definitions commonly refer to the cost associated with adjustments (Fankhauser 1998) and development initiatives (World Bank 2010) required to reestablish farming conditions prevailing before the occurrence of variations in climate while the IPCC (2014) consider the cost associated with planning, preparing for, facilitating, and implementing adaptation measures in farming practices including transaction costs. The definitions illustrate that there is no consensus as to what constitute adaptation cost. There are various aspects being considered by different authors and organizations. Another challenge is that there is no distinction according to type and/or class of adaptation costs. This makes the issue a complex phenomenon.

In this chapter, some components of the above definitions were adapted. Other considerations mentioned by Smith and Ward (1998) and Meyer et al. (2013) while assessing the costs of natural hazards were also infused. As such, in this chapter, intangible adaptation cost refers to problems and dangers that maize farmers encounter while planning, preparing, and implementing adjustments that cushion the impact of climate variability shocks on maize farming. The problems and dangers are not measurable in monetary terms, as they are not traded on the market as Smith and Ward (1998) illustrated. On the other hand, indirect adaptation cost refers to secondary unintended effects that unfold during implementation of adjustments to cushion the effect of climate variability. The secondary unintended effects include interruptions of normal day-to-day operations, extra demand on available resources such as labor and post adaptation effects that may arise. These may be measurable but not necessarily in monetary terms and are secondary effects of adaptation.

Climate justice scholarship on adaptation raises questions of fairness (Adger et al. 2013; Forsyth 2014). To achieve fairness, intangible and indirect costs arising from the implementation of adaptation measures especially among social classes within societies should be recognized. This chapter builds on the concept of "fair adaptation" (Adger et al. 2013; Forsyth 2014; Mikulewicz 2017) and draws upon components of distributive and procedural fairness (Graham et al. 2018) to explore intangible and indirect costs arising from the implementation of adaptation measures among farmers in Chirumanzu. Emphasis is on one of the four principles of fair adaptation, "putting the most vulnerable first" in order to ensure equitable outcomes among those at risk (Paavola and Adger 2006). Thus, concern is on redressing existing inequalities and preventing future ones (Graham et al. 2015) through prioritizing vulnerable resettlement farmers in Chirumanzu District of Zimbabwe and other smallholder farmers elsewhere in similar settings.

Of particular interest to this chapter is the fact that existing adaptation costs have been largely direct and tangible (Meyer et al. 2013). Apart from that, adaptation cost literature has concentrated on attaching cost figures (Mundial 2006; United Nations Framework Convention on Climate Change (UNFCCC) 2008; World Bank 2010) to existing direct and tangible costs yet there are other costs difficult to measure in quantitative terms. This implies that literature on adaptation cost to date does not fully enhance understanding of the overall picture of the cost associated with adaptation. This problem makes it difficult for smallholder farmers to make sustainable decisions and adopt and maintain adaptation plans apart from diminishing adaptive capacity and resilience. Yet governments, researchers, and international organizations are increasingly encouraging adaptation among smallholder farmers.

This chapter, therefore, responds to the necessity of understanding the intangible and indirect components of the cost of adaptation to climate variability. Intangible and indirect costs associated with adaptation to climate variability for smallholder maize farmers in resettlement areas of Chirumanzu District of Zimbabwe were explored. To achieve this, problems and dangers associated with adaptation measures arising from implementing adaptation plans are established. Subsequently, intangible and indirect aspects are discovered. Three important questions are answered in this chapter: What adaptation measures did maize farmers in resettlement areas of Chirumanzu commonly adopt? What problems and dangers are associated with adaptation measures commonly adopted? What are the intangible and indirect costs associated with the problems and dangers?

Location and Climate of Chirumanzu Resettlement Areas and Rationale for Selection

Figure 1 shows the location of Chirumanzu in Zimbabwe and the resettlement areas in the district. The District lies between longitudes 29°50E and 30°45E and latitudes 19°30S and 20°20S. Chirumanzu District is located in the Midlands Province of Zimbabwe. At least 90% of Chirumanzu District lie in Natural Region III while the remainder falls under Natural Region IV (Gwamuri et al. 2012). Natural Region III receive rainfall ranging from 500 to 750 mm, while Natural Region IV receive 400–510 mm per annum (Musara et al. 2011).

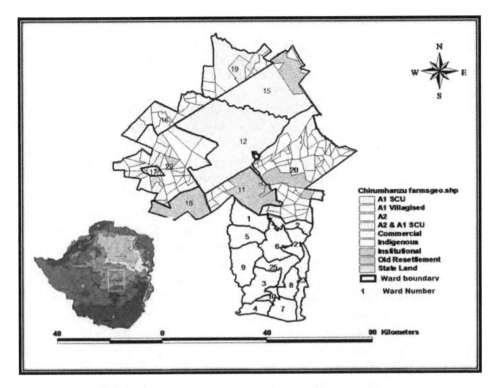

Fig. 1 Map of Chirumanzu District, Zimbabwe showing resettlement areas

Chirumanzu resettlement area experiences extreme weather events in the form of severe mid-season dry spells and frequent seasonal droughts (Simba and Chayangira 2017) yet rain fed agriculture is the major source of livelihood in the area with maize farming being the major farming activity as it is the staple food. Maize farming is largely for consumption. Surplus is often sold providing a source of income for the resettlement farmers. Such a rural setting that is continually battered by climate variations and extreme events presented the need for farmers in the resettlement areas to adapt making them suitable candidates for the study thus presenting a platform for interrogation.

Of the 23 wards in Chirumanzu District, nine (Wards 11, 12, 15, 16, 17, 18, 19, 20, and 22) are predominantly resettlement areas. These resettlement areas were a result of both the old resettlement program in the 1980s and the Fast Track Land Reform Program (FTLRP) of 2000. Farmers were resettled under A1 and A2 models. This chapter focuses on beneficiaries of the A1 model because these are local communities where the effects of climate variability are largely felt as recognized by the IPCC (2014). Model A1 was designed to address poverty and vulnerability for the landless poor. Furthermore, it was expected that by doing so, the congested communal areas would be depopulated resulting in relatively small farms that could sustain families (UNDP 2002).

Model A1 has three settlement schemes namely villagized, self-contained (Njaya and Mazuru 2014), and old resettlement. In the villagized scheme, each farmer is allocated about one hectare to build homesteads in a village set up. Each resettled farmer got 5–6 hectares away from the village. Grazing area is designated to be communal. In contrast to the villagized scheme, self-contained plots ranging from 15 to 30 hectares are allocated per farmer for both cultivation and grazing. Old resettlement scheme is similar to the villagized scheme in its setup. As Thebe (2018) reveals, A1 farms were established on former ranching farms with varying land quality characterized by poor sandy soils to rich black loams. Chirumanzu resettlement areas lie over four main soil types. These are deep sandy, clay, shallow sodic, and sandy loam. These soil types are more or less similar to those found in most smallholder farming communities across Zimbabwe and Southern Africa. Therefore, the information provided in this chapter is of greater applicability to most smallholder farmers.

To explore the intangible and indirect costs of adaptation, four out of the nine resettlement wards were selected. The enquiry was therefore, conducted in Wards 11, 12, 15, and 20. Dominant soil types and resettlement schemes in Chirumanzu District were considered specifically to capture intangible and indirect adaptation cost experiences of different maize farmers operating under different circumstances. There are four dominant soil types in Chirumanzu resettlement areas and three A1 resettlement schemes. Each ward represented a specific soil type and resettlement scheme.

Wards 11, 12, 15, and 20 represented sandy loam, shallow sodic, clay, and deep sandy soils, respectively. Apart from this consideration, Ward 11 represented the old resettlement scheme and farmers with relatively more farming experience but with relatively small farm sizes. Ward 12 was selected to represent the villagized scheme and farmers with less farming experience with relatively small farm sizes. Ward 15 was selected to represent self-contained scheme and farmers with less farming experience with relatively larger farm sizes. Ward 20 was unique because it has both the A1 villagized and self-contained resettlement schemes. Furthermore, there is need to note that the imbalance in the main soil types and resettlement models justified inclusion of Ward 20 in the investigation.

Characteristics of A1 Maize Farmers in Chirumanzu Resettlement Areas

Fifty-four A1 maize farmers were identified from the selected wards so that they could serve as the sources of data through semi-structured interviews. The inclusion and/or exclusion criterion that suited the theme of the investigation was A1 maize farmers who have adapted to climate variability and still had operational adaptation systems in place during the time of the study. Intensive consultation with the District Agricultural Extension Officer and Ward Extension Officers led to the identification of A1 maize farmers who met the inclusion/exclusion criteria.

Out of the 54 farmers, 10 were female. Farmers' age showed skewed outcomes. Forty-three farmers were in the 61–70 and 71–80 age groups with only two being 31–40 years old. Farming experience varied from six to more than 30 years. Thirty-one farmers attained secondary education. Only three farmers had tertiary qualifications. Five farmers did not have any formal education but could read and write. Most of the farmers were settled during the period 1998 and 2002 implying that they were settled under the FTLRP. All farmers in Ward 11 were settled in the 1980s reflecting that they were settled under the old resettlement scheme. Only seven farmers were settled in the 1990s and beyond 2002. Farmers settled under the Fast Track Land Reform were either under villagized or self-contained schemes. Farm sizes for farmers under A1 old resettlement and A1 villagized schemes were either five or six hectares while those in the self-contained scheme were either 15 or 30 hectares. Thirty-five farmers had at least 6 hectares of arable land while more than 50 farmers had between 3 and 10 hectares of arable land under maize production.

Data Gathering

Semi-structured interviews were conducted with the selected A1 maize farmers. Interviews were conducted in Shona, which is the local vernacular language. This ensured that the respondents had a common understanding of the meaning of the questions compared to the situation had English been used. A semi-structured interview guide with open-ended questions was used to gather data on commonly adopted adaptation measures, associated problems and dangers, as well as the related intangible and indirect costs. Detailed notes of the interviews were taken. Concurrent audio recording of the interview proceedings helped enhance accuracy of farmers' responses.

Data saturation was reached between the 8th and 9th farmer in all the wards. Instead of terminating the interviews, they were continued until the 15th farmer in Wards 11, 12, and 20. This was done in line with the Peterson (2019) advice that seeks to obtain deeper insights. However, in Ward 15 interviews were terminated after interviewing the 9th farmer. An unexpected commotion developed during interviewing, which created hostile conditions that made it impossible to continue with the data collection as originally planned. In total, 54 farmers were interviewed. This final sample was decided on based on the Morse (2016) recommendation that in a grounded theory research such as this one, 30–50 interviews should be conducted. Four additional farmers were included based on the assumption that new and rich data could be generated from them. In addition, Onwuegbuzie and Leech (2007) argue that it is important to ensure that a sample is neither too small to achieve data saturation nor too big to manage.

Data Handling and Analysis

All audio-recorded interviews were first transcribed verbatim. Textual data from audio recordings and notes taken were stored as a MS Excel spreadsheet on a case-based entry as illustrated by Friese (2016). The file was imported into Atlas.ti Version 8. A grounded theory approach was adopted during the thematic content analysis carried out in Atlas.ti Version 8. Inductive thematic content analysis was performed through reading responses given by the farmers. Textual responses were used to develop preliminary codes through inductive coding. It was performed via open and in vivo coding, in line with the Friese (2016) method. Open coding involved reading the text responses, sentence by sentence while forming detailed and structured themes. In this way, a grounded analysis was guaranteed. Simultaneously, codes and resulting code groups that were drawn from primary data were certified while avoiding missing important data. The same approach was used for in vivo coding. In this case, a word or phrase from textual responses was used to represent a code or code group.

Similar or related codes with the same meanings were merged to avoid unnecessary repetition. Irrelevant codes were deleted. Preliminary codes were grouped and merged into code groups. Groups with preliminary codes that were combined yet reflecting two or more concepts were split. Selective coding was used to create qualitative visual representations of the data in the form of network diagrams. Relationships and patterns were created using the resulting codes and groups linking them with quotations to create network diagrams, which were then exported to MS Word for use in presenting results and research report.

Adaptation Measures Adopted by Maize Farmers in Resettlement Areas of Chirumanzu

Maize farmers in resettlement areas of Chirumanzu adopted six common measures to cushion the impact of climate variability shocks on maize farming. The nature of the adaptation measures suggests that maize farmers in resettlement areas commonly adopt autonomous, *ex-post* measures (Smit et al. 2000) in response to climate variability shocks and impacts. The act of adaptation is done after farmers have already experienced significant impact costs. This signifies that adaptation is an act of restoration among maize farmers in Chirumanzu rather than intentional. Farmers adopted measures out of desperation in order to restore the losses incurred due to the impact of climate variability. This corresponds to Shoko et al. (2016) who compared adoption rates and preference of adaptation measures among smallholder farmers and found out that for some measures, adoption rates were low while preference was high and vice versa. It was therefore concluded that often farmers adopt adaptation measures not because they prefer them but out of desperation.

It is noticeable that some measures were highly adopted in some wards than in others possibly due to different characteristics of farmers. Of particular importance to this chapter are the soil types and resettlement schemes. Deviant and exceptional cases, adopted by only a few or even one farmer, were also observed among the measures adopted by maize farmers. Furthermore, some measures were adopted for specific stages in the maize value chain, while some were adopted for more than one production stage. Apart from that, some measures were adopted to address several climatic variations.

In some cases, some of the measures adopted were contradictory, while others ended up disadvantaging farmers leaving them in a worse off situation. It is also important to note that out of the six adaptation measures, female farmers adopted five which is a competitive number. This outcome challenges the binary male-female view of gender that women are passive victims of climate change (Nellemann et al. 2011) and confirms that women are proactive agents when it comes to climate adaptation (Mitchell et al. 2007; Dankelman 2010). In the following section, the main adaptation measures that maize farmers in resettlement areas of Chirumanzu commonly adopt are described and discussed.

Changing Planting Dates

Maize farmers in resettlement areas change planting dates in various ways. Farmers either plant early (dry planting), well before the rains start, or late as they are forced to wait for effective rains (sometimes late December or even January). Early and/or dry planting is adopted to address late onset of rains while avoiding delayed planting and falling behind production schedule. Late planting while waiting for effective rains is also normally adopted to address late onset of rains while avoiding poor germination rates and poor crop stand. In some cases, farmers plant their maize crop at different dates of the farming season (staggering). Staggering is adopted to address recurring climate variability shocks and evade total crop failure. Early and late planting are contradictory measures suggesting that adapting to climate variability is not "a one size fits all" approach. Different farmers take different routes depending on background characteristics bringing out farmer heterogeneity.

Use of Drought Tolerant Varieties

The use of drought tolerant varieties is one of the highly adopted measures in resettlement areas of Chirumanzu. Drought tolerant varieties increases yield in most drought stricken areas by an average of 600 kgs per hectare (Lunduka et al. 2019). Drought tolerant varieties endure moisture stress for a period of six weeks and have high tolerance to dry spells especially during the critical stages of development (Cairns et al. 2013). Due to recurring droughts that are experienced almost every three years in resettlement areas of Chirumanzu, farmers use drought tolerant varieties to improve yields. Drought is one of the limiting factors in rain-fed maize farming especially in sub-Saharan Africa particularly in Zimbabwe (Lunduka et al. 2019). This explains why the use of drought tolerant varieties is one of the highly adopted measures in resettlement areas of Chirumanzu. Common drought tolerant varieties grown by maize framers in resettlement areas of Chirumanzu are hybrid varieties, SC513 and SC403.

    Although at times a considerable proportion of farmers receive drought tolerant varieties from the government, some are always not fortunate enough to get access. As such, despite most of the resettled farmers being resource poor (Mushunje et al. 2003; Chinamatira et al. 2016) and confronted with a variety of challenges in acquiring inputs including hybridized seed, Mkodzongi and Lawrence (2019) opined that resettlement farmers strive to self-finance farming activities. This can be argued to include adaptation investments such as the use of drought tolerant varieties. In that case, the use of drought tolerant varieties is an act of desperation that farmers implemented in order to restore and or prevent the losses from the impact of climate variability.

Good Crop Establishment Practices (No Regret Measures)

Good crop establishment practices were mainly "no regret measures." These measures include effective and timeous weeding, timeous application of fertilizers, timeous harvesting, and irrigation. According to Hallegatte (2009), no-regret measures are measures that yield benefits even if climate variability does not occur. Similarly, good crop establishment practices are normal practices conducted during maize farming whether there is climate variability or not. Such measures illustrate manipulative behavior and it is argued that they are a subset of adaptive behavior (Thomsen et al. 2012). Good crop establishment practices such as timeous weeding are an example of measures adopted for specific production stages in the maize value chain. Timeous weeding was adopted solely for weed control stage. On the other hand, irrigation is one of the exceptional cases adopted by only a few farmers in resettlement areas of Chirumanzu.

Conservation Farming

Conservation farming is mainly practiced through the use of planting basins and mulching in resettlement areas of Chirumanzu. These adaptation measures are mainly adopted to conserve moisture while at the same time reducing moisture stress. Planting basins is the use of shallow pits that allow accumulation of water, facilitating rapid infiltration into the soil (Rusinamhonzi 2015). The use of planting basins is highly debated in literature. In Zimbabwe, it is referred to as *"dhiga ufe"* translated as the "dig and die" technology (Andersson et al. 2011) due to the high labor requirements associated with the practice. Mulching is the use of crop residue and/or other organic material to maintain a permanent or semi-permanent soil cover with the intention of conserving soil moisture (Nyamangara et al. 2014). Maize farmers in resettlement areas of Chirumanzu often use grass and old leaves since crop residue is not adequate to cover large areas of land. Conservation farming is common in Ward 11 with sandy loam soils that quickly loses soil moisture due to accelerated drying of sandy loam soils.

Wetland Farming

Maize farmers in Chirumanzu often move from their original farms to the wetland areas due to recurring droughts, mid-season dry spells, unpredictable, and unreliable rains. In these wetlands, they can plant as early as September because the soil has enough moisture to sustain germination. Wetland farming is an irregular measure that is an unusual feature in existing research. The adaptation measure is also a characteristic of certain places that have wetland areas, for example, Ward 20 of Chirumanzu. Maize farmers in Ward 20 relocated to the wetland areas in the same locality as their original farms. This is not fully consistent with the conservative narrative on disaster-induced migration (Gray and Mueller 2012). The conservative narrative predicts that climate variability consistently increase long-term population mobility and effects are most visible for long distance moves. However, in Chirumanzu Ward 20, effects are most visible for internal moves. It also important to note that wetland farming is a form of maladaptation. According to the IPCC (2014), maladaptation refers to actions that increases the risk of adverse climate-related outcomes, increases vulnerability to climate variability, or diminishes welfare, now or in the future. The concept of maladaptation as Magnan et al. (2016) puts it focuses on the importance of accounting for potential side effects of adaptation to avoid solutions that are worse than the original problem. Moving to the wetland shifts environmental pressures elsewhere and is thus considered a form of maladaptation (Magnan et al. 2016).

Diversification

Diversification was practiced to avoid the effects of continual decline in maize yield and total crop failure. It is practiced in three different ways. These are crop, enterprise, and income diversification. Farmers grow multiple crops such as small grains, groundnuts that are more resilient to extreme climate variability than maize. Farmers venture into new enterprises like broiler production and look for alternative activities such as gold panning as alternative sources of income than relying on maize alone. Diversification is adopted by farmers with larger farms mostly in the self-contained scheme with more than 15 hectares of land. This illustrate findings by Amare et al. (2018) who projected that farm size has a significant and positive effect on the adoption of diversification to combat climate change impacts.

Problems and Dangers Associated with Adaptation Measures Adopted by Maize Farmers

Maize farmers in resettlement areas of Chirumanzu experienced several problems and dangers while implementing adaptation measures. Figure 2 shows an imported network diagram depicting a visual representation of the problems and dangers associated with adaptation measures adopted by maize farmers. The network diagram displays code groups (key problems and dangers) and codes (associated problems and dangers). Seven code groups and 32 codes were comprehended. This implies that 32 problems and dangers were identified.

Code groups are presented in red boxes on the network diagram and associated codes in white boxes. Associated codes are linked to code groups with arrows showing how they are related to respective code groups.

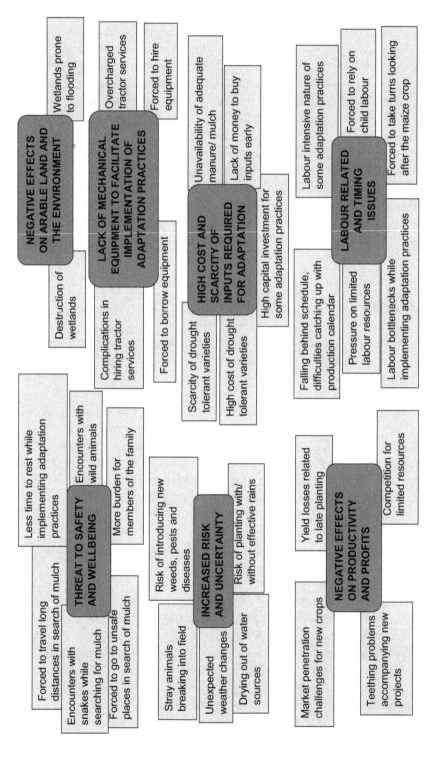

Fig. 2 Network diagram showing problems and dangers associated with adaptation measures adopted by A1 maize farmers

Threat to Safety and Wellbeing

Some of the adaptation measures implemented by maize farmers pose threat to safety and wellbeing of farmers. Tedious and laborious adaptation measures such as conservation farming result in less time to rest. Mulching forces farmers to travel long distances in search of mulch material. Sometimes they were forced to go to unsafe places in search of grass and old leaves which they normally use as mulch material. Encounters with snakes while searching for mulch material are common.

Lack of Mechanical Equipment

Lack of mechanical equipment that facilitates implementation of adaptation practices is a major problem. Implements such as ploughs, picks, spades, and hand hoes are required to facilitate implementation of adaptation plans. However, due to resource constraints, farmers do not always have these tools at hand when needed. Farmers are forced to either borrow or hire equipment to execute adaptation plans. However, there is no guarantee that when they borrow they will get the implements. On the other hand, the process of hiring equipment is complicated and this leads to delays and interruptions in implementing adaptation activities. Tractor services are scarce in the area and a few who have them overcharged the services.

High Cost and Scarcity of Inputs

High cost, scarcity, and continual increase in prices of inputs is also a major problem for farmers while adapting to climate variability. Inputs required to implement or maintain some adaptation practices are not readily available. Where these inputs were found, usually on the black market, they are often charged double the normal price. Scarcity and high cost of inputs hinders farmers from implementing adaptation plans on time.

Increased Risk and Uncertainty

Farmers encounter increased risk and uncertainty while adapting to climate variability. Unexpected weather changes ruin adaptation plans in general. Planting early pose a risk of stray animals breaking into the field since during September when most people in the area start early planting, people will still be sending their cattle away without someone attending to them. Planting without effective rains often leads to poor germination posing the risk of seed rotting and subsequently seed wastage. Irrigation is associated with the risk of drying out of water sources.

Labor Related and Timing Issues

Labor related and timing issues is another key problem that farmers encounter during adaptation. Waiting for effective rains results in delayed land preparation subsequently leading to delayed planting. Resultantly, farmers fall behind schedule. Sometimes they plant late December or January. This puts pressure on labor resources as they try to catch up with production calendar. The effectiveness of their work is compromised as they do things hurriedly. Some adaptation measures are labor intensive. The use of planting basins requires a lot of labor. Such measures burden farmers and they end up relying on children to assist. Furthermore, such measures are limited by labor bottlenecks.

Negative Effects on Productivity and Profits

Some adaptation measures call for extra production costs. This threatens the sustainability of maize enterprises. Where they practice wetland farming, they plant as early as mid-September. Farmers will need to buy fence to create a barrier in case stray animals might enter into the field and eat maize while they are not watching. Equally, diversification demands extra investment costs to start new project ventures. Farmers ventured into the production of horticultural crops, broiler production, and production of small grains. However, these production activities present a new set of challenges for farmers. Teething problems are experienced as farmers start these new project ventures. Competition for resources is escalated since there is need to balance the resources for all the activities. Farmers who diversify into small grains face marketing strategies as there is no established market for small grains.

Intangible Costs Associated with Adaptation to Climate Variability

Problems and dangers encountered by maize farmers while adapting to climate variability have several intangible costs. Intangible costs do not have a market value therefore cannot be valued in monetary terms. Tables 1 and 2 summarize adaptation measures adopted by maize farmers, associated problems and dangers, as well as the related intangible costs.

Health, Wellbeing, and Safety Concerns

Some adaptation measures threaten farmers' wellbeing raising health and safety concerns. Due to the lack of mulch material, farmers and household members are forced to travel long distances and go to unsafe places in search of adequate and suitable materials. As such, there is less time to rest and exposure to danger due to the risk of encountering snakes. With wetland farming, farmers are forced to take

Table 1 Intangible and indirect costs associated with adaptation measures adopted by A1 maize farmers

Adaptation measure	Problems and dangers associated with adaptation measures	Associated intangible costs	Associated indirect costs
Conservation farming	Tedious and labor intensive	Reliance on child labor leading to violation of children's rights	Family labor opportunity cost
		Extra burden on members of the family	
	Lack of mechanical equipment	Setbacks in implementing	Loss due to delays
	Forced to borrow equipment	Embarrassment associated with borrowing	–
		Ridicule and stereotyping associated with borrowing	–
		Availability and access of equipment not guaranteed	–
	Forced to hire equipment	Effort put in hiring equipment	Unplanned, additional hiring cost
		Availability of equipment not guaranteed	–
	Complications in hiring equipment	Effort put in negotiating deals	Opportunity cost of lost time negotiating
	Overcharged services for equipment	–	Extra, often unplanned cost
	Lack of mulch material	Setbacks in implementing	–
	Forced to travel long distances	Less time to rest	–
		Wellbeing concerns	–
	Forced go to unsafe places	Encounters with snakes	–
		Safety concerns	–
	Possibility of introducing new weeds, pests, and diseases with mulch	–	Yield loss related to the risk of new weeds, pests and diseases

(continued)

Table 1 (continued)

Adaptation measure	Problems and dangers associated with adaptation measures	Associated intangible costs	Associated indirect costs
Changing planting dates	Risk of falling behind schedule and difficulties catching up while waiting for effective rains	Worry, anxiety and uncertainty	Yield losses related to timing
		Extra burden	Opportunity cost of labor
	Unexpected weather changes	Pain and suffering due to losses	Yield losses related to unexpected weather changes
	Possibility of replanting associated with dry planting as seed fail to germinate due to insufficient moisture	Pain and suffering due to poor germination	Losses due to seed wastage
		Threatened emotional wellbeing	–
		–	Cost of additional seed for replanting
		–	Additional labor cost for replanting
	Late planting is associated with yield losses	–	Yield loss related to late planting

turns in looking after the maize crop in case stray animals might break into the field and destroy the crops. While looking after the maize crop, they encounter wild animals such as wild pigs and baboons which may attack them.

Burden on Household Members

Adapting to climate variability exert excessive burden on household members. Conservation farming, for example, is generally labor intensive, tedious, and time consuming. Farmers usually rely on family members particularly children and women to provide the required labor as they cannot afford to hire or contract workers.

As a result, children get to school late, often miss school, and have little time to play as they are expected to help with implementing adaptation activities. The female folk are excessively burdened as farmers implement adaptation activities. Women are forced to disregard other activities such as going to church. Other important activities like tending to the cattle and goats are left to the children.

Looking after and taking care of the home is put on hold or sometimes neglected. Women face difficulties in coping and balancing household and adaptation chores. Household members struggle with abnormal day schedules and unusual working

Table 2 Intangible and indirect costs associated with adaptation measures adopted by A1 maize farmers

Adaptation measure	Problems and dangers associated with adaptation measures	Associated intangible costs	Associated indirect costs
Use of drought tolerant varieties	High cost of drought tolerant varieties	Effort spent looking for better priced DTVs	Opportunity cost of time
	Scarcity of drought tolerant varieties	Effort spent looking for DTVs	
Good crop establishment practices	Tedious nature of practices such as effective weeding	–	Investment cost for efficient system
	Time consuming	–	–
	High startup costs associated with irrigation	–	Investment cost for irrigation system
	Risk of drying up of water sources	Uncertainty and worry over water source	
Diversification	Capital investment for new project	–	Investment cost for new project
	Teething problems while starting a new project	Pain and suffering as project fail to take up	Losses due to failure of new project
	Competition for resources	Indecision/difficulties allocating resources	–
	Market penetration challenges for new products	Difficulties penetrating the market	–
Wetland farming	Risk of stray animals breaking into field, Need to take turns to look after maize	Taking turns looking after maize, forced to forego other activities	Opportunity cost of activities foregone
	Encounters with wild animals	Possibility of attacks from wild animals, threat to safety	–
	Wetlands prone to flooding and waterlogging	Difficulties accessing wetlands	–

hours as they sometimes spent the whole day in the field and sometimes wake up as early as 3 o'clock in the morning. Priority is given to adaptation activities more than other household activities. This shows that adaptation activities disproportionately burden the vulnerable (Barnett and O'Neill 2010) particularly women and children. Reliance on family labor increases vulnerability of women and children. Adaptation in resettlement areas of Chirumanzu is therefore not socially equitable for women and children (Barnett and O'Neill 2010).

Worry, Anxiety, and Uncertainty

Farmers fall behind schedule when they change planting dates to wait for effective rains. Farmers worry and feel anxious as they wait with uncertainty for rains to come. Farmers feel helpless as they wait with uncertainty as to when the effective rains will come. Farmers who irrigate also worry over the risk of drying out of water sources.

Pain and Suffering

Farmers experience pain and suffering due to the poor germination rates and poor crop stand due to insufficient moisture associated with changing planting dates through early and/or dry planting. They are distressed over the inputs, resources, and effort wasted. After early and/or dry planting sometimes very little rains come and they result in seed rot thereby wasting seed, fertilizers, and labor. Farmers also experience pain and suffering as new project ventures fail to start up progressively when they diversify.

Effort Spent to Facilitate Smooth Implementation of Adaptation Measures

The high cost and scarcity of drought tolerant varieties force farmers to spend a lot of time going from one place to the next, sometimes from one town to the next looking for better priced seed. Drought tolerant varieties are no longer readily available on the market in Zimbabwe. Since 2009, the traditional suppliers of drought tolerant varieties have been failing to meet demand because they were unable to cope with hyperinflation (Dekker and Kinsey 2011). Since then drought tolerant varieties have been scarce on the formal market only covering less than 50% of the demand (Willems 2014). This facilitated the mushrooming of a parallel market for drought tolerant varieties where they are charged double or more. Incidentally, this forces farmers to make effort looking for better prices drought tolerant varieties they can afford.

Indecision During Allocation of Scarce Resources

Farmers experience difficulties in allocating scarce resources as they diversify into new project ventures. Potential conflicts in labor allocation and capital investment sharing are common to farmers who diversify. Indecision is therefore common while allocating resources as all the projects will be important to them.

Ridicule and Embarrassment Associated with Borrowing Equipment

The lack of mechanical equipment forces farmers to borrow. When they borrow the implements, they are subjected to ridicule and experience feelings of embarrassment. This shows that support networks for adaptation practices sparsely exist among maize farmers in Chirumanzu. Instead of offering social support to other farmers while implementing adaptation practices (Townsend et al. 2015), fellow community members ridicule them. Cooperation and solidarity which are important mechanisms determining the extent to which adaptation measures are adopted are nonexistent among maize farmers in resettlement areas.

The above narrative can be drawn back to the way in which the resettlement program was structured. In particular, the A1 model, communities were created overnight and by chance since most of the times farms were allocated by picking out a number from a hat (Chiweshe 2014). This led to the establishment of "stranger neighboring households" (Barr 2004: 1753) who did not know one another and were therefore forced by to settle and interact. As such, different groups of people with competing views, opinions, and interest would create a conducive environment for noncooperation (Chiweshe 2014).

Indirect Costs of Adaptation to Climate Variability

Indirect costs are secondary unintended effects of adaptation activities. Although measurable, they are not valued directly on the market. They are unintentional and often unplanned effects of adaptation with a time lag. Tables 1 and 2 summarize adaptation measures adopted by maize farmers, associated problems and dangers, as well as the related indirect costs.

Family Labor Opportunity Cost

The use of family labor to implement adaptation measures raises the issue of family labor opportunity costs. Although family labor opportunity cost has been identified in existing literature, the indirect aspects surrounding it are not well articulated. In existing literature, labor is considered a variable cost that is direct and tangible as is measurable in man-days per hectare. However, despite the existence of several family labor opportunity cost valuation measures, a common unit of measurements that specifies the cost of important activities foregone while providing labor for adaptation plans is nonexistent.

Maize farmers in resettlement areas rely mostly on family labor. In most cases, farm owners, household members including women and children are forced to forego other important activities while providing labor for adaptation plans. All members of the household provide the required labor for adaptation activities. Commonly, the labor hours provided by household members go unnoticed and mostly unpaid. Hence, the net value of time spent in the next best activity would have been foregone. Major effects are on children who are excessively burdened as they are obliged to help before they go to school and after school. As such, adaptation disproportionately burdens the most vulnerable (Eriksen and O'Brien 2007) groups of children in resettlement areas which constitute unsustainable adaptation.

Existing literature on family labor does not go further to establish potential conflicts in labor allocation between adaptation through measures such as planting basins and other household duties. Yet, Rusinamhonzi (2015) indicated that there are indeed potential conflicts in labor allocation between adaptation through planting basins and other activities. This tally with findings of this study where adaptation practices were often drawn back by labor bottlenecks. This study illustrate that the use of planting basins disproportionately burdened maize farmers and their families while reducing the incentive to adapt (Barnett and O'Neill 2010). This explains why very few farmers in Chirumanzu adopted it. Potential conflicts in labor allocation among adaptation plans and competing household duties are a form of adaptation cost that is unnoticed.

Additional and Unplanned Cost of Production

Maize farmers are confronted with additional often unplanned costs while implementing adaptation plans. Adaptation measures that require special equipment result in maize farmers incurring unplanned hiring costs. Farmers are constrained by the lack of necessary implements such as hoes, ploughs, and tractors to execute adaptation plans. Farmers resort to hiring implements at a cost paid either paid in cash or in kind. This increase the cost of production and most farmers cannot afford it. Wetland farming calls for extra production costs since farmers need to buy fence to create a barrier that obstructs stray animals from entering the field and eat the maize while they are not watching. This relates to the contention of Adger et al. (2009) and Morrison and Pickering (2012) that inadequate technology presents additional unplanned costs for farmers while executing adaptation plans. Payments made in kind are not given much thought; hence, the cost associated usually goes undetected. Early and/or dry planting often results in replanting twice or thrice when seed fail to germinate due to climate variability extremes such as little rains. This generates additional seed, fertilizer, and chemical requirements per hectare thus raising total input cost. These costs are usually misconstrued as normal variable costs without considering the indirect, additional, and unplanned costs associated.

Yield Losses Related to Risk, Unavoidable Delays, and Timing

Some adaptation measures increased the likelihood of loses rather than gains (Pittelkow et al. 2014). For example, late planting results in grain yield loses of up to 5% for each week of delayed planting (Nyagumbo 2008). In Zimbabwe, Nyakudya and Stoosnijder (2015) mentioned incidences of pests and diseases as the other reasons why late planting often give lower yields. Similarly, this study established that farmers who diversified with small grains ended up failing to get a market for this alternative crop. Nonetheless, since literature on adaptation largely focuses on benefits brought forth by adaptation and neglects the costs, these processes are rarely deliberated in adaptation cost assessments.

Huge Investment Cost for Some Adaptation Practices

Diversification demands huge investment costs to start new enterprises. Equally, irrigation requires high start-up costs to secure the water source pump and other irrigation facilities for an efficient system. Maize farmers are resource constrained as such these costs are enormous for them.

Conclusion

Investing in measures that reduce the impact of climatic variation in maize farming is indeed the missing link for A1 farmers in resettlement areas of Chirumanzu. Measures such as conservation farming, use of drought tolerant varieties, changing planting dates, practicing good crop establishment, wetland farming, and diversification may strengthen maize farmers to fulfill their massive role of safeguarding food and nutrition security. The Government of Zimbabwe intends to make adaptation a national priority. On the international front, adaptation has been identified as the only solution for furthering the Climate Change Policy. This has seen an extensive adoption of various adaptation measures around the globe particularly among smallholder farmers including A1 farmers in Zimbabwe. Several success stories on adaptation have been widely documented. Despite considerable variations in effectiveness of adaptation measures in maize farming, yields have been reportedly improved. Overall, adaptation reduces vulnerability and improves resilience while at the same time reducing incidences of rural poverty. However, overall adaptive capacity and resilience among smallholder farmers especially in the African region including Zimbabwe is reportedly still low. This is partly due to the problems and dangers that accompany the process of adaptation. As such, caution must be taken especially in communities that are making efforts to adapt so that the problems and dangers are managed accordingly. The problems and dangers originate from planning, executing, monitoring, and maintenance of adaptation systems. The problems and dangers are associated with some intangible costs that are difficult to measure in quantitative terms and therefore challenging to assign a monetary value which makes them repeatedly ignored in adaptation cost assessments. The problems and dangers are also associated with indirect costs that are secondary effects of adaptation and cannot be easily comprehended at face value as they are not directly measured in monetary terms. This chapter brings to light the fact that in some cases, adaptation results in problems and dangers that limit adaptive capacity and increase vulnerability to some extent. This chapter contributes to existing literature and argues that adaptation does not only bring positive outcomes. The chapter progresses the argument that intangible and indirect costs are an enormous part in adaptation planning and cost assessments. This chapter advocates for prioritization of the intangible and indirect costs associated with the problems, dangers, and unintended adaptation effects of adaptation to increase uptake and enhance sustainability. Stakeholders in the climate adaptation arena should not overlook intangible and indirect costs associated to the problems, dangers, and unintended effects that come with adaptation activities in order to enhance social and environmental justice. It is time to practice what is being preached in the climate adaptation arena and not neglect crucial prerequisites of adaptation such as the "first do no harm principle."

References

Adam B, Wiredu AN (2015) Cost benefit analysis of climate change adaptation measures on soil and water conservation in Northern Ghana (eds: Ministry of Food and Agriculture GUOH, Institute of Farm Management (Germany)). Csir-Savanna Agricultural Research Institute, Ghana. Online, Ministry of Food and Agriculture, Republic of Ghana

Adger WN, Dessai S, Goulden M, Hulme M, Lorenzoni I, Nelson DR, Naess LO, Wolf J, Wreford A (2009) Are there social limits to adaptation to climate change? Clim Chang 93:335–354

Adger W, Tara Q, Lorenzoni I, Murphy C, Sweeney J (2013) Changing social contracts in climate-change adaptation. Nat Clim Chang 3:330–333

Agrawala S, Crick F, Jette-Nantel S, Tepes A (2008) Empirical estimates of adaptation costs and benefits: a critical assessment. In: Agrawala S, Fankhauser S (eds) Economic aspects of adaptation to climate change: costs, benefits and policy instruments. OECD, Paris

Agrawala S, Bosello F, Carraro C, de Cian E, Lanzi E (2011) Adapting to climate change: costs, benefits and modelling approaches. Int Rev Environ Resour Econ 2011(5):245–284

Amare ZY, Ayoade JO, Adelekan IO, Zeleke MT (2018) Barriers to and determinants of the choice of crop management strategies to combat climate change in Dejen District, Nile Basin of Ethiopia. Agric Food Secur 7(37)

Andersson J, Giller K, Mafongoya P, Mapfumo P (2011) Diga udye or diga ufe? (Dig-and-eat or dig-and-die?) Is conservation farming contributing to agricultural involution in Zimbabwe? Proceedings of the Regional Conservation Agriculture Symposium, 8–10 February 2011, Emperor's Palace Hotel and Conference Centre, Johannesburg

Arfunuzzaman MD, Mamnun N, Islam S, Dilshad T, Syed A (2016) Evaluation of adaptation practices in the agricultural sector of Bangladesh: an ecosystem based assessment. Climate 4:11

Barnett J, O'Neill S (2010) Maladaptation. Glob Environ Change Hum Policy Dimens 20:211–213

Barr A (2004) Forging effective new communities: the evolution of civil society in Zimbabwean resettlement villages. World Dev 32(10):1753–1766

Basset TJ, Fogeliman C (2013) Deja vu or something new? The adaptation concept in climate change literature. Geoforum 48(2013):42–53

Biagini B, Bierbaum R, Stults M, Dobardzic S, McNeely SM (2014) A typology of adaptation actions: a global look at climate adaptation actions financed through the Global Environmental Facility. Glob Environ Chang 25(2014):97–108

Cairns JE, Crossa J, Zaidi PH, Grudloyma P, Sanchez C, Araus JL, Thaitad S, Makumbi D, Magorokosho C, Bänziger M, Menkir A (2013) Identification of drought, heat, and combined drought and heat tolerant donors in maize. Crop Sci 53(4):1335–1346

Chambwera MA, Heal G, Dubeux C, Hallegatte S, Leclerc L, Markandya A, Neumann J (2014) Economics of adaptation. In: Climate change 2014: impacts, adaptation, and vulnerability. Part A: Global and sectoral aspects. Contribution of Working Group II to the Fifth Assessment

Report of the Intergovernmental Panel on Climate Change. Cambridge University Press, Cambridge, UK/New York

Chinamatira L, Mtetwa S, Nyamadzawo G (2016) Causes of wildland fires, associated socio-economic impacts and challenges with policing, in Chakari resettlement area, Kadoma, Zimbabwe. Fire Sci Rev 5(1):1

Chiweshe MK (2014) Understanding social and solidarity economy in emergent communities: lessons from post-fast track land reform farms in Mazowe, Zimbabwe. UNRISD Occasional Paper: Potential and Limits of Social and Solidarity Economy, No 1. United Nations Research Institute for Social Development (UNRISD), Geneva

Costinot A, Donaldson D, Smith C (2016) Evolving comparative advantage and the impact of climate change in agricultural markets: evidence from 1.7 million fields around the world. J Polit Econ 124(1):205–248

Dankelman I (2010) Gender and climate change: an introduction. Routledge, United Kingdom

Dekker M, Kinsey B (2011) Coping with Zimbabwe's economic crisis: small-scale farmers and livelihoods under stress. African Studies Centre, Working Paper 93/2011, Leiden

Doczi J, Ross I (2014) The economics of climate change adaptation in Africa's water sector. A review and way forward. Overseas Development Institute Report Working paper. Overseas Development Institute, London

Eriksen SH, O'Brien K (2007) Vulnerability, poverty and the need for sustainable adaptation measures. Clim Pol 7(2007):337–352

Fankhauser S (1998) The cost of adapting to climate change. Working paper no 16. Global Environmental Facility, Washington, DC

Fankhauser S (2009) The costs of adaptation to climate change. Centre for Climate Change Economics and Policy. Paper no 8/Working paper no 7. Grantham Research Institute on Climate Change and the Environment, London

Forsyth T (2014) Climate justice is not just ice. Geoforum 54:230–232

Friese S (2016) CAQDAS and grounded theory analysis. Max Planck Institute for the Study of Religious and Ethnic Diversity, Göttingen

Graham S, Barnett J, Fincher R, Mortreux C, Hurlimann A (2015) Towards fair local outcomes in adaptation to sea-level rise. Climatic Change 130(3):411–424

Graham S, Barnett J, Moertreux C, Hurlimann A, Fincher R (2018) Local values and fairness in climate change adaptation: insights from rural Austrailian communities. World Dev 108 (2018):332–343

Gray CL, Mueller V (2012) Natural disasters and population mobility in Bangladesh. Proc Natl Acad Sci 109(16):6000–6005

Gwamuri J, Mvumi B, Maguranyanga EF, Nyagumbo I (2012) Impact of Jatropha on rural livelihoods case of Mutoko District, Zimbabwe. RAEIN-Africa Secretariat, Windhoek

Hallegatte S (2009) Strategies to adapt to an uncertain climate change. Glob Environ Chang 19 (2009):240–247

Intergovernmental Panel on Climate Change (2014) Climate change 2014: impacts, adaptation and vulnerability. Part A: Global and sectoral aspects. In: Contribution of working group II to the Fifth assessment report of the intergovernmental Panel on Climate Change (eds: Field CB, Barros VR). Cambridge University Press, New York

Kumar K, Shyamsundar P, Nambi AA (2010) The economics of climate change adaptation in India–research and policy challenges ahead. Policy Note 6:10–42

Lunduka RW, Mateva KI, Magorokosho C, Manjeru P (2019) Impact of adoption of drought-tolerant maize varieties on total maize production in south Eastern Zimbabwe. Clim Dev 11 (1):35–46

Magnan AK, Schipper ELF, Burkett M, Bharwani S, Burton I, Eriksen S, Gemenne F, Schaar J, Ziervogel G (2016) Addressing the risk of maladaptation to climate change. Wiley Interdiscip Rev Clim Chang 7(5):646–665

Mccarl BA (2007) Adaptation options for agriculture, forestry and fisheries. A Report to the UNFCCC Secretariat Financial and Technical Support Division. http://agecon2.tamu.edu/people/faculty/mccarl-bruce/papers

Menike L, Arachchi K (2016) Adaptation to climate change by smallholder farmers in rural communities: evidence from Sri Lanka. Proc Food Sci 6:288–292

Meyer V, Becker N, Markantonis V, Schwarze R, Van Den Bergh J, Bouwer L, Bubeck P, Ciavola P, Genovese E, Green CH (2013) Assessing the costs of natural hazards-state of the art and knowledge gaps. Nat Hazards Earth Syst Sci 13(5):1351–1373

Mikulewicz M (2017) Politicizing vulnerability and adaptation: on the need to democratize local responses to climate impacts in developing countries. Clim Dev 10(1):18–34

Milman A, Arsano Y (2014) Climate adaptation: contradictions for human security in Gambella, Ethiopia. Glob Environ Chang 29(2014):349–359

Mitchell T, Tanner T, Lussier K (2007) 'We know what we need': South Asian women speak out on climate change adaptation, ActionAid and Institute of Development Studies (IDS), University of Sussex

Mkodzongi G, Lawrence P (2019) The fast-track land reform and agrarian change in Zimbabwe. Rev Afr Polit Econ 46(159):1–13

Morrison C, Pickering C (2012) Limits to climate change adaptation: case study of the Australian Alps. Geogr Res 51(1):11–25

Morse JM (2016) Mixed method design: Principles and procedures (Volume. 4). Routledge

Mundial B (2006) Investment framework for clean energy and development. J Sports Sci, World Bank, Washington, DC

Musara JP, Zivenge E, Chagwiza G, Chimvuramahwe J, Dube P (2011) Determinants of smallholder cotton contract farming participation in a recovering economy: empirical results from Patchway district, Zimbabwe. J Sustain Dev Afr 13(4):1–3

Mushunje A, Belete A, Fraser GC (2003) Technical efficiency of resettlement farmers of Zimbabwe. Proceedings of the 41st Annual Conference of the Agricultural Economic Association of South Africa (AEASA), 2–3 October 2003, Pretoria

Nellemann C, Verma R, Hislop L (2011) Women at the frontline of climate change: gender risks and hopes. A rapid response assessment. United Nations Environment Programme, GRID-Arendal. UNEP, Arendal

Njaya T, Mazuru N (2014) Emerging new farming practices and their impact on the management of woodlots in A1 resettlement areas of Mashonaland Central Province in Zimbabwe. Asian Dev Policy Rev 2(1):1–19

Nyagumbo I (2008) A review of experiences and developments towards Conservation Agriculture (CA) and related systems in Zimbabwe. In: Goddard T, Zoebisch MA, Gan YT, Ellis W, Watson A, Sombatpanit S (eds) No-till farming systems. World Association of Soil and Water Conservation, Bangkok

Nyakudya IW, Stoosnijder L (2015) Conservation tillage of rain fed maize in semi-arid Zimbabwe: a review. Soil Tillage Res 145(2015):184–197

Nyamangara J, Mashingaidze N, Masvaya EN, Nyengerai K, Kunzekweguta M, Tirivavi R, Mazvimavi K (2014) Weed growth and labor demand under hand-hoe based reduced tillage in smallholder farmers' fields in Zimbabwe. Agric Ecosyst Environ 187:146–154

Onwuegbuzie AJ, Leech NL (2007) A call for qualitative power analyses. Qual Quant 41 (1):105–121

Paavola J, Adger WN (2006) Fair adaptation to climate change. Ecol Econ 56(4):594–609

Peterson JS (2019) Presenting a qualitative study: a reviewer's perspective. Gift Child Q 63 (3):147–158

Pittelkow CM, Liang X, Linquist BA, van Groenigen KJ, Lee J, Lundy ME, van Gestel N, Six J, Venterea RT, van Kessel C (2014) Productivity limits and potentials of the principles of conservation agriculture. Nature 517:365–368

Rurinda J, Mapfumo P, van Wijk MT, Mtambanengwe F, Rufino MC, Chikowo R, Giller KE (2013) Managing soil fertility to adapt to rainfall variability in smallholder cropping systems in Zimbabwe. Field Crop Res 154:211–225

Rusinamhonzi L (2015) Tinkering the periphery: labour burden not crop productivity increased under no-till planting basins on smallholder farms in Murehwa district, Zimbabwe. Field Crop Res 170(2015):66–75

Sachikonye LM (2003) From 'growth with equity' to 'fast-track' reform: Zimbabwe's land question. Rev Afr Polit Econ 30(96):227–240

Schipper ELF (2009) Meeting at the crossroads? Exploring the linkages between climate change and disaster risk reduction. Clim Change Dev 1(1):16–30

Shoko D, Oloo G, Francis J, Kori E (2016) Strategy preference vs adoption: a comparative analysis for Agro-Small and Medium Enterprises' survival in a challenging economic environment. The Dyke: Journal of the Midlands State University (Special Issue):150–169

Simba FM, Chayangira J (2017) Rainfall seasons analysis as a guiding tool to smallholder farmers in the face of climate change in Midlands in Zimbabwe. J Earth Sci Clim Chang 8(3)

Smit B, Burton I, Klein RJ, Wandel J (2000) An anatomy of adaptation to climate change and variability. In: Societal adaptation to climate variability and change. Springer, Dordrecht

Smith K, Ward R (1998) Floods: physical processes and human impacts. Wiley, New Jersey

Thebe V (2018) Youth, agriculture and land reform in Zimbabwe: experiences from a communal area and resettlement scheme in semi-arid Matabeleland, Zimbabwe. Afr Stud 77(3):336–353

Thomsen DC, Smith TF, Keys N (2012) Adaptation or manipulation? Unpacking climate change response strategy. Ecol Soc 17(3)

Townsend I, Awosoga O, Kulig J, Fan H (2015) Social cohesion and resilience across communities that have experienced a disaster. Nat Hazards 76(2015):913–938

United Nations Development Programme (2002) UNDP interim mission report. Harare

United Nations Framework Convention on Climate Change (2008) Investment and financial flows to address climate change: an update. United Nations Framework Convention on Climate Change Report. United Nations, New York

Willems F (2014) Access to inputs in Zimbabwe: changes since the Fast Track Land Reform Programme. Dissertation submitted in partial fulfilment of Masters in Human Geography, Globalisation, Migration and Development at Radbound University

World Bank (2010) Economics of adaptation to climate change: synthesis report. The World Bank Group, Washington, DC

Permissions

The contributors of this book come from diverse backgrounds, making this book a tru
international effort. This book will bring forth new frontiers with its revolutionizing resear
information and detailed analysis of the nascent developments around the world.

We would like to thank all the contributing authors for lending their expertise to make t
book truly unique. They have played a crucial role in the development of this book. Witho
their invaluable contributions this book wouldn't have been possible. They have made vi
efforts to compile up to date information on the varied aspects of this subject to make th
book a valuable addition to the collection of many professionals and students.

This book was conceptualized with the vision of imparting up-to-date information a
advanced data in this field. To ensure the same, a matchless editorial board was set t
Every individual on the board went through rigorous rounds of assessment to prove the
worth. After which they invested a large part of their time researching and compiling t
most relevant data for our readers.

The editorial board has been involved in producing this book since its inception. They ha
spent rigorous hours researching and exploring the diverse topics which have resulted
the successful publishing of this book. They have passed on their knowledge of decad
through this book. To expedite this challenging task, the publisher supported the team
every step. A small team of assistant editors was also appointed to further simplify t
editing procedure and attain best results for the readers.

Apart from the editorial board, the designing team has also invested a significant amou
of their time in understanding the subject and creating the most relevant covers. Th
scrutinized every image to scout for the most suitable representation of the subject a
create an appropriate cover for the book.

The publishing team has been an ardent support to the editorial, designing and productio
team. Their endless efforts to recruit the best for this project, has resulted in t
accomplishment of this book. They are a veteran in the field of academics and their po
of knowledge is as vast as their experience in printing. Their expertise and guidance h
proved useful at every step. Their uncompromising quality standards have made this bo
an exceptional effort. Their encouragement from time to time has been an inspiration f
everyone.

The publisher and the editorial board hope that this book will prove to be a valuable pie
of knowledge for researchers, students, practitioners and scholars across the globe.

List of Contributors

endai Nciizah
epartment of Sociology, Rhodes University,
akhanda, South Africa

inah Nciizah
epartment of Development Studies,
vishavane Campus, Midlands State
niversity, Zvishavane, Zimbabwe

aroline Mubekaphi
hool of Agricultural, Earth and
nvironmental Sciences, University of
waZulu-Natal, Scottsville, South Africa

dornis D. Nciizah
il Science, Agricultural Research Council –
stitute for Soil, Climate and Water, Pretoria,
uth Africa

aac Rutenberg, Arthur Gwagwa and
Ielissa Omino
PIT, Strathmore University, Nairobi, Kenya

lasade Mary Owoade
epartment of Crop Production and Soil
ience, Ladoke Akintola University of
chnology, Ogbomoso, Nigeria

muel Godfried Kwasi Adiku
epartment of Soil Science, University of
hana, Legon, Ghana

hristopher John Atkinson
atural Resources Institute, University of
reenwich, London, UK
epartment of Agriculture, Health and
nvironment, Natural Resources Institute,
niversity of Greenwich, Chatham, UK

ilys Sefakor MacCarthy
il and Irrigation Research Centre Kpong,
niversity of Ghana, Accra, Ghana

Irene Ojuok
National Technical specialist Environment
and Climate Change, World Vision Kenya,
Nairobi, Kenya

Tharcisse Ndayizigiye
SMHI/Swedish Meteorological and
Hydrological Institute, Nairobi, Kenya

David Karienye
Department of Geography, Garissa University,
Garissa, Kenya

Joseph Macharia
Department of Geography, Kenyatta
University, Nairobi, Kenya

Alima Dajuma
Department of Meteorology and Climate
Sciences,West African Science Service Centre
on Climate Change and Adapted Land Use
(WASCAL), Federal University of Technology
Akure (FUTA), Ondo State, Nigeria
Laboratoire de Physique de l'Atmosphère
et de Mécaniques des Fluides (LAPA-MF),
Université Félix Houphouët-Boigny, Abidjan,
Côte d'Ivoire

Siélé Silué
Université Peleforo Gon Coulibaly, Korhogo,
Côte d'Ivoire

Kehinde O. Ogunjobi
Department of Meteorology and Climate
Sciences,West African Science Service Centre
on Climate Change and Adapted Land Use
(WASCAL), Federal University of Technology
Akure (FUTA), Ondo State, Nigeria
Federal University of Technology Akure
(FUTA), Ondo State, Nigeria

Heike Vogel and Bernhard Vogel
Institute of Meteorology and Climate Research, Karlsruhe Institute of Technology (KIT), Karlsruhe, Germany

Evelyne Touré N'Datchoh and Véronique Yoboué
Laboratoire de Physique de l'Atmosphère et de Mécaniques des Fluides (LAPA-MF), Université Félix Houphouët-Boigny, Abidjan, Côte d'Ivoire

Arona Diedhiou
Laboratoire de Physique de l'Atmosphère et de Mécaniques des Fluides (LAPA-MF), Université Félix Houphouët-Boigny, Abidjan, Côte d'Ivoire
Université Grenoble Alpes, IRD, Grenoble INP, IGE, Grenoble, France

David O. Yawson
Centre for Resource Management and Environmental Studies, The University of the West Indies, Bridgetown, Barbados

Michael O. Adu and Paul A. Asare
Department of Crop Science, University of Cape Coast, Cape Coast, Ghana

Frederick A. Armah
Department of Environmental Science, University of Cape Coast, Cape Coast, Ghana

Angeline Mujeyi and Maxwell Mudhara
College of Agricultural, Engineering and Science, Discipline of Agricultural Economics, University of KwaZulu-Natal, Scottsville, Pietermaritzburg, South Africa

Miftah F. Kedir
WGCFNR, Hawassa University, Shashemene, Ethiopia

Central Ethiopia Environment and Fore Research Center, Addis Ababa, Ethiopia

George Olanrewaju Ige, Oluwole Matthe Akinnagbe and Olalekan Olamigo Odefadehan
Department of Agricultural Extension ar Communication Technology, School Agriculture and Agricultural Technolog Federal University of Technology, Akur Nigeria

Opeyemi Peter Ogunbusuyi
Department of Agricultural and Resour Economics, School of Agriculture ar Agricultural Technology, Federal Universi of Technology, Akure, Nigeria

Oluwatobi Ololade Ife-Adediran
Geochronology Division, CSIR-Nation Geophysical Research Institute (NGR Hyderabad, India
Department of Physics, Federal Universi of Technology Akure, Akure, Ondo Stat Nigeria

Oluyemi Bright Aboyewa
Department of Physics, College of Arts ar Sciences, Creighton University, Omaha, N USA

Saumu Ibrahim Mwasha and Zoe Robinsc
School of Geography, Geology and tl Environment, Keele University, Staffordshir UK

Dumisani Shoko Kori, Joseph Francis ar Jethro Zuwarimwe
Institute for Rural Development, Universi of Venda, Thohoyandou, South Africa

ndex

aptive Capacity, 3, 5, 51, 72-75, 77, 81-82, 97, 107-108, 116, 119, 121, 123-124, 142, 157, 7, 210, 227

ing, 10

ricultural Production, 2, 4, 15, 73-74, 76, 78, 80, 8, 161, 165, 195, 198

nual Rainfall, 6, 77

onymity, 192

thropogenic Activities, 87, 93, 97

able Crops, 168, 171

odiversity, 24, 75, 79, 108, 177, 184

rbohydrate, 129

rbon Dioxide, 73, 150, 176

ssava, 41-45, 47-49, 163, 191, 198

real Crop, 129

mate Change, 1-9, 11, 13-18, 20-24, 26-28, -39, 51, 53-56, 58-61, 67-68, 71, 122-131, 136, 0-142, 144-145, 155-162, 164, 167-168, 171-9, 192-195, 210, 215, 217, 227-231

mate Variability, 38, 68, 72-82, 101, 129, 161, 5, 173-174, 177-178, 188-189, 194, 196, 198-1, 206-212, 214-217, 219-220, 222, 225-226, 1

mate Variation, 174, 179

mate-smart Agriculture, 16, 38, 128-130, 141-142, 4

matic Conditions, 3-5, 7, 59, 75, 177, 179

matic Situations, 76

nservation Agriculture, 8, 15, 17, 80, 130-131, 1-142, 204, 228, 230

op Failure, 2, 10, 75, 79, 173, 203, 215, 217

op Productivity, 15, 41, 49-50, 52, 76, 231

eforestation, 24, 56, 60, 63, 75, 101, 144-145, 6, 159

esertification, 56

sease Control, 200

Droughts, 2-4, 6-10, 21, 26, 30, 55, 74-75, 78, 104, 123, 161, 211, 215-216

Dry Season, 77, 87-88, 101, 105, 111, 117, 119, 136, 163, 191

Dry Spells, 6-7, 9, 199, 211, 215-216

E
Ecological Zones, 18, 40-43, 47, 49-50, 189-192, 201, 203

Ecosystem, 58, 64, 75, 103, 107-110, 116-119, 121, 124-127, 176, 204, 228

Environmental Degradation, 75-76, 104, 117, 123

Extreme Temperatures, 75-76

F
Farming Practices, 198-199, 209, 230

Farming Systems, 52, 128-129, 131, 139, 142, 230

Floods, 3, 21-22, 26, 30, 55, 68, 74-75, 79, 104, 111, 116, 119, 123, 161, 199, 231

Food Availability, 76

Food Crop, 162, 174

Food Insecurity, 4, 6, 8, 17, 54, 60, 72-76, 80, 104, 108, 117, 119, 123, 203, 208

Food Security, 2-3, 8-9, 14-18, 24, 33, 36-38, 56, 61-62, 71-72, 76-77, 103, 106, 108-110, 117-119, 122, 124-127, 141, 163, 173-174, 188, 202-204

G
Geometry, 180

Germplasm, 66

H
Heat Stress, 9, 74, 116, 120

Herbicides, 129

Horticulture, 125, 136, 198

Humidity, 112, 163

I
Intercropping, 15, 49, 130-131, 136-137, 170, 172, 199

L
Land Degradation, 53-54, 56, 59, 75-76, 79

Legumes, 17, 131

Livestock, 6-8, 13, 54-56, 59-65, 67, 74-76, 80, 82, 87, 89, 98, 128-129, 131-133, 135, 137, 140-141, 161, 191, 200-201, 203

M

Maize Crop, 6, 128, 132, 215, 218, 222

Meteorology, 84-85, 89, 91-92

Methane, 73

Methionine, 10

Minerals, 10, 41, 129

N

Natural Disasters, 29, 58, 229

Nitrous Oxide, 73

Nutrition, 9, 17, 19, 41, 51, 54, 62-63, 82, 126, 204, 207-208, 227

O

Organic Agriculture, 15

Organic Matter, 41, 49-51, 145, 158, 199-200, 204

Oxygen, 144, 149

P

Pesticide, 132, 135, 137

Pesticides, 118, 129, 202

Phosphorus, 10, 41, 45, 50, 129

Phytochemicals, 142

Plant Species, 76, 152-153

Precipitation, 3, 6-8, 31, 35, 55, 75-76, 79, 82, 87-89, 92-93, 98, 100, 161, 177-178, 182, 184-185

R

Rainy Season, 7, 77-78, 92

Rationale, 206, 210

Renewable Energy, 30, 98, 176-177, 179-180, 183, 185

Rural Population, 54

S

Sea Level, 146, 177, 181, 191

Semiarid Lands, 72-74, 77

Small-scale Farmers, 12, 14, 229

Smallholder Farmers, 2-7, 9-12, 15-16, 18, 81, 128-130, 141, 161, 203, 208, 210, 212, 214, 22 230-231

Soil Erosion, 56, 77, 131, 161, 198, 205

Soil Water, 51, 198

Sustainability, 3, 10, 17, 28, 51, 105-109, 115, 125, 127, 142, 145, 159, 175-177, 182-183, 22C 228

Sustainable Agriculture, 17, 50, 81

Sustainable Development, 27-28, 36, 38, 51, 72 75, 115, 186

T

Topography, 29

Transmission, 60, 115

Tree Planting, 81, 198

Tropical Regions, 87

V

Vegetation Cover, 49, 75, 77

W

Weather Patterns, 7, 21, 35, 77-78, 81

Wet Season, 43, 100, 105

Printed in the USA
CPSIA information can be obtained
at www.ICGtesting.com
JSHW062236071123
51533JS00031B/55

9 781639 8769